THOMAS E. S   80

LET ME BE ONE OF THEM

THE WORLD'S FIRST SPACESHIP

SHUTTLE

Stackpole Books

THE WORLD'S FIRST SPACESHIP

SHUTTLE

ROBERT M. POWERS

SHUTTLE: THE WORLD'S FIRST SPACESHIP

Published by
STACKPOLE BOOKS
Cameron and Kelker Streets
P.O. Box 1831
Harrisburg, Pa. 17105

Published simultaneously in Don Mills, Ontario, Canada by Thomas Nelson & Sons, Ltd.

Jacket art courtesy of NASA.

Printed in the U.S.A.

Library of Congress Cataloging in Publication Data

Powers, Robert M.   1942–
    Shuttle: the world's first spaceship.

    Bibliography: p.
    Includes index.
    1. Reusable space vehicles.   I. Title.
TL795.5.P68 1979   629.45'4   79–652
ISBN 0–8117–1686–4
ISBN 0–8117–2112–4 pbk.

To Dr. Robert Goddard, who was content to get his hands dirty, to work and research ten years in trying to make a rocket go up just one mile, and to the men and women who will fly in the space shuttle and those who will come after them

The achievements of the United States space program have not only widened the range of human knowledge, but have also served to unite all Americans and, indeed, all nations of the world in a common adventure of discovery.

Senator Edward M. Kennedy

We are on a journey to keep an appointment with whatever we are.

Gene Roddenberry, producer of ''Star Trek''

Foreword by Michael Collins   11

Preface   13

Acknowledgments   17

1   Night Launch from Cape Canaveral   19

2   The Brick Moon and Other Fancies   35

3   Appointment in Orbit   47

4   Aboard an Orbiter   61

5   Launches, Landings, and In Between   85

6   The New Astronauts   105

7   The World of Space Shuttle   119

8   Spacelab   141

CONTENTS

9   Astronomy in Space   157

10   Modules, Space Platforms, and Space Stations   171

11   The Industrialization of Space   181

12   Solar Power Satellites   197

13   Beyond the Space Shuttle   211

     Appendix A: NASA Abbreviations   231

     Appendix B: Space Shuttle Diagrams   233

     Glossary   245

     Additional Readings   250

     Index   252

FOREWORD

The idea behind the shuttle is simple enough: to design a spacecraft which can be used over and over again, unlike present rockets which fall into the ocean at the end of their first and only voyage. But reusability is no simple matter. As this book quickly makes apparent, the shuttle has to do so many things so well, that its designers have created a machine of incredible complexity. The shuttle is designed to be a workhorse, and it looks it. About the size of a DC-9 but not nearly as sleek, to me the shuttle looks like a cross between the Concorde and a Mack truck. It takes off vertically like a rocket, lands horizontally like an airplane, and between the two will be responsible for delivering to orbit our country's space payloads for the next several decades.

When my friend John Young first rockets into orbit aboard the shuttle, he will be doing a lot of things differently than he did during his Gemini and Apollo flights. Once the two gigantic solid rocket boosters ignite, the shuttle is utterly dependent on them, with no escape rockets to pull it free in case of danger. On the first flights, the two-man crew will be provided ejection seats, but when the crew grows to four or seven, the ejection seats will be removed, and from then on the solids simply must not fail.

Nor will there be any unmanned flight test of the shuttle. The Titan II rocket which powered the Gemini had had dozens of test flights before men were permitted to ride it, and the Saturn V moon rocket had two unmanned flights before Frank Borman and his crew rode the third to the moon on Christmas Eve of 1968. But the shuttle is ushering in a new era, and John Young will be out to prove that flying in space is no longer an experiment, but a routine operation.

John will also find reentry into the earth's atmosphere a lot different. Up until now, spacecraft have always come into the atmosphere backwards, with the heat shield on their blunt end providing protection until it was time to open the parachutes and plop gently into the sea. Not so with the shuttle: it is the world's most sophisticated glider. Starting its orbital descent near Australia, its speed will gradually decrease from 18,000 mph to 200 mph as it reaches its runway in the California desert. John will be helped by five onboard computers as he delicately maneuvers down through the hypersonic, supersonic, and subsonic flight regimes, along the way facing every uncertainty known to aerodynamicists since Kitty Hawk.

To a pilot or scientist, the shuttle brings exciting times, but even more important, it opens doors to dreamers as well. Robert Powers knows this, and in this book he skillfully blends the proven technology of the shuttle with the promise of the stars. He makes our future in space seem real, and sensible, and I think he is right.

MICHAEL COLLINS

PREFACE

Like a Phoenix, Project Apollo rose from the crushed dreams, fire, and ashes of disaster, where three astronauts died, to *Apollo 11* and triumph in the Sea of Tranquility. Men landed on the moon, tested its surface, brought back its rocks—the more for us to wonder about—and saw the earth rise from a cratered terrain on an alien world and float blue and insignificant in the black space above. It was a historic time, a gigantic and great endeavor realized. Its impact will be with us for a very long time.

Since the *Apollo* flights, we have continued our probing and fishing about out there. On Venus, strange shale rocks lurk among the footpads of a Russian vehicle. Atmospheric probes and American Venus orbiters fly to the second planet. Out there in the deeps of space, *Voyager* coasts toward the giant bulk of Jupiter, and *Pioneer* drifts onward to an encounter with the planet Saturn and its rings of rock and ice.

Project Viking ticks slowly away, two years after it landed on Mars, the Red Planet, that place we have so often in this century thought of as a sister world. In a few years, an orbiter will go to Jupiter. It will swing over the gas giant for a year, two years, or more, to investigate the atmosphere and the curiosities of its many moons. Perhaps we will find life there in that primordial soup which is the atmosphere of Jupiter, or on Saturn's large moon, Titan, or elsewhere in this small and lonely corner of the universe, far from the center of the Milky Way galaxy.

We have embarked on a grand adventure. Less than eight thousand years ago we could not write our own language. Eight thousand years ago, only a tiny fraction of earth's history, we could not properly provide for or protect our children. We could not prevent our own early deaths from diseases unimagined.

In a few short years, because of the space shuttle, a tiny fraction—perhaps two thousands—of our race will have been to the space beyond the earth. Before this century's end, there will be people living in space year round, and some few of us alive at this moment will live to see the first birth of a child high above the planet where we live. And that child may see the first child born on another planet. The generation count will not be long before one of their descendents will watch the first starship leave.

But none of this will happen if we do not continue to develop our civilization in space.

If we do not keep our thrust into space moving at a high rate, it is not we who now live on earth who will suffer. It will be those who will come after, who may have to live in a world which space technology could have saved and was not allowed to. It may be those who live in a country that did not choose to recognize its own potential at the moment in history when it should have.

Space shuttle, with an appropriate program, is the key to the technology of the next century. It is the key to a sprawling tumble of outward expansion to meet the unknowns of the universe. It is a key to planning missions to Venus to bring back those strange rocks, flying to Mars to ponder making it into a planet fit for mankind, driving deep into the emptiness near Pluto, and landing once again on the moon.

Shuttle is also the key to turning our space technology inward to cope with the results of our own selfishness and destructiveness. With shuttle, we can monitor the earth on remote sensing devices planted in space. We can service satellites to improve communications the world over, perfect navigation satellites, and partially solve the energy problems with solar-powered satellites.

And somewhere, too, in all our hopes and designs for shuttle, is the thin optimism that somehow mankind will deal with space differently—above all, more elegantly and sensibly—than we have with similar affairs before.

If we choose to do so, if we are willing to grasp the opportunity, if we influence our politicians, we will see the first permanent space station above the earth, with men and women working, dreaming, and learning the potential of a totally new frontier. Looking further, there may be colonies in space, cities in the sky against the background of the stars.

We may huddle beneath the surface of the long-dead moon or perhaps dwell on the surface in huts. We may dare the long leap to Mars and actually walk under the hurtling moons of Barsoom, or probe the murk and smog of Venus to understand that planet and what it may tell us about ourselves and the way we have behaved on our own planet.

The atmospheres and surfaces of the planets farther away than Venus and Mars will come under our space cameras and be spread before our instruments and computers as data bits. The first rocks from Mars, the first whiffs of atmosphere from distant Titan, and the first soil from Venus may be ours before the century's end.

This, above all others, is a time in which history will be made. The human race will make another giant leap of conquest. We should not lose this chance, or worse, ignore its implications. It would be nice if the end of it could be seen, but that is always denied us. At least, we will see the beginning, and that will have to be enough.

ACKNOWLEDGMENTS

For helping to provide information on the space shuttle, the author would like to thank Dan Robertson of Martin Marietta Aerospace; Dick Barton of Rockwell International Space Division; Dave Garrett, Hugh Harris, and Bob Gordon of the National Aeronautics and Space Administration. My special thanks to Les Gaver, director of NASA Washington Audio Visual Services, for letting me go through their files and for providing many of the illustrations used in the book.

I owe a special debt to Neil McAleer who patiently listened to the hundreds of ideas which I poured forth on how the book should be done and who made suggestions on how it could be done better. Betty Chancellor and Alison Brown worked long and hard editing the manuscript.

To the hundreds and thousands of scientists, engineers, dreamers, writers, politicians, and administrators who laid the groundwork so that I could be alive to see the threshold of space breached and a foothold established, I can only say that I would not be alive as happily at any other moment in history unless it would be during the first manned landing on Mars—and that I also expect to see.

We are now living the stories that were dreamed by science-fiction dreamers not more than twenty, thirty, forty years ago.

Leonard Nimoy

1

NIGHT LAUNCH FROM CAPE CANAVERAL

It is a clear night, with only a few clouds in the northwest toward Daytona Beach and little patches of fog hovering over the Atlantic. People in Titusville, across the Indian River, are tuning in Johnny Carson or getting ready for bed. The wind coming in from the sea, pushing the fog patches closer and closer to the Florida coast, is chilly but not cold.

Venus, now an evening star, set several hours ago, and Jupiter is a handful of degrees above the eastern horizon. The stars spread out above the glow of lights surrounding launch pad 39A at Kennedy Space Center. Along the perimeter road are a circle of searchlights, their beams overlapping as they try to cover three hundred feet of vertical distance from the pad's concrete base to the top of the space shuttle.

This is the spot on earth where five Saturn V engines raised a 363-foot (111-meter) rocket which hurled a 50-ton manned *Apollo* capsule to the moon in 1969, leaving the names Armstrong, Collins, and Aldrin forever in history and in the minds of millions of people. It was from this pad that *Skylab* was launched above the blue and white planet to become a space station visited by three crews of astronauts before it was left in a slowly decaying orbit. It was from here that a mission began that culminated in the joining of the hands of two countries nominally cool to each other—*Apollo-Soyuz*.

19

In launch control at Kennedy Space Center, the consoles flicker in the faces of a dozen key technicians. A score of monitors are watched by anxious men whose sole responsibility is to see that they tell the story everyone is trained to hear: "GO." Out on the press site, television crews smoke and wait, and a few people speak softly into tape recorders. A dozen cameras on tripods are lined up at the white spaceship poised beneath an autumn Florida sky.

The crew of the space shuttle arrives, without bulky space suits: seven small, ordinary people, dwarfed by a giant aluminum dragon that will shortly breathe a great splash of fire and throw them up into the space above the earth. The seven get on the pad elevator for the ride up to a room in the service tower—the white room. The shuttle is ready for flight except for fueling. High above the pad, in the elevator, the crew listens to the creaks and groans of the vehicle.

With "Snoopy" helmets on, the crew moves from the white room to the shuttle orbiter. The pilot commander and his copilot are the first through the hatch and climb up into the forward seats of the flight deck. The cockpit is a jungle of controls and gauges; red and green lights flicker on the panels, pointers behind glass covers move back and forth, and cathode-ray tubes in color screens like small TV sets shift symbols almost too quickly for the eye to follow.

Two mission specialists enter through the round hatch behind the pilot and copilot and continue up to the two seats on the aft flight deck. Using a ladder, the remainder of the crew climbs to three seats in the mid-deck, just forward of the air-lock compartment. Everyone straps in—two shoulder belts, a belt across the lap, and a crotch strap. The crew technician helps them plug into the command and communications circuit so the crew is heard live through the Snoopy helmets.

The ladder used in the mid-deck for crew entry is removed, and the technician ducks out of the hatch with a wave. The handles on the hatch spin, and the shuttle orbiter with the crew inside is sealed from the outside world. The cabin atmosphere smells faintly antiseptic. The long wait of prelaunch begins.

The service towers roll clear of the shuttle. The white room where the crew made last-minute checks and said good-bye is already many feet away from the spaceship. From their vantage point high in the air of the orbiter suspended on the external skin of the fuel tank, the crew sees the lights of Cape Canaveral glow past the floodlights in the distance and the waves ride over the sand on the hotel beaches of Cocoa.

While the crew waits, the fuel loading starts. First, liquid hydrogen is loaded into the rear of the giant external tank. As the fuel load comes on, the structure begins a song of groans and clanks, creaks, and assorted other dangerous-sounding but harmless noises. The intense cold of the liquid hydrogen shrinks the aluminum in the shuttle's fuel tank and in other parts of the ship. As the fuel pours in, the contracting metals respond with shrieks. Outside the insulated orbiter, the liquid hydrogen turns to gas and hisses and boils off through the vents in the external tank.

An artist's view of a space shuttle liftoff from Cape Canaveral. Shuttle's main engines and solid rocket boosters are firing. (Courtesy Rockwell International Space Division)

The liquid oxygen comes in silently, despite its cold, for the external tank is chilled now by the liquid hydrogen and it will not contract. At T-minus-three minutes, the liquid oxygen tank is topped up. The final topping up of the liquid hydrogen tank is completed, replacing the fuel which has boiled off through the vents during filling. At T-minus-two, the liquid oxygen tank is pressurized. At T-minus-one minute, the hydrogen fuel is pressurized. At Mission Control, the situation is as tense as it always is with one minute to go on a launch. It has been called, by pilots and ground alike, the world's longest minute—about "an hour" long.

"*All system go for launch.*" Mission Control is ready.

"Roger. All go here." The shuttle is ready.

The agonizing drama of the countdown begins. "Nineteen . . . eighteen . . . seventeen . . ." (The idea of a reverse countdown was not the invention of some rocket engineer, nor was it invented by NASA. It was invented by someone with a relatively loose association with space research and rocketry, director Fritz Lang, in 1929. He needed a device to give high drama to his movie, *Frau im Monde,* and invented the countdown.)

At ten seconds, the flickering lights on the orbiter's instrument panel go to green. Across the upper flight deck, there are steady banks of solid green. The turbo-pumps start up with a growl.

"Nine . . ." comes from Mission Control over the earphone. It echoes around the press site from loud speakers and inside the buildings west of pad 39A.

BAM!

"We have SSME ignition."

"Roger."

The thrust from the three shuttle main engines builds for a few seconds, sucking thousands of gallons of liquid oxygen and liquid hydrogen from the external tank. The orbiter vibrates and the noise is a summer thunderstorm magnified.

"Five . . . four . . ."

The solid rocket boosters come on with a jolt, kicking the thrust of the entire vehicle to almost seven million pounds.

"We have SRB ignition."

"Roger."

"Liftoff!" The sound from Mission Control is tinny and far away in the presence of the surroundsound of the launch thunder. "We have liftoff."

"Roger, we copy."

The launch tower is cleared at T-plus-six seconds, and by thirty seconds after launch the shuttle is traveling faster than a Concorde. Eleven tons of fuel a second

Shuttle roars up toward near-earth orbit. (Courtesy NASA)

The SRBs, solid rocket boosters, separate from the space shuttle at twenty-seven nautical miles altitude. Using small thrusters fore and aft, the SRBs will move away from the shuttle and then impact at a preselected point in the ocean. They will be recovered by ships for reuse on later space shuttle flights. (Courtesy Rockwell International Space Division)

are going into the shuttle main engines and being burned up by the solid rocket boosters. From below, the only thing visible in the dark night sky is the flame.

A little after seven seconds into the flight, the control of the space shuttle is shifted to Houston control, at Johnson Space Center. At T-plus-one minute, Houston control comes on to announce thrust reduction. The acceleration drops off to stay within tolerable range. There are none of the bone-jarring, mind-shattering vibrations and strap-to-strap buffetings of the *Apollo* Saturn V beast. This is a smooth take-off, hardly 3-g at maximum thrust.

T-plus-two minutes and the earth is rapidly falling away. Outside the windows the sky is a deep, dark blue. Indigo turns to black as the shuttle climbs up, 100 miles (160.9 kilometers) downrange from the Cape. Speed is now 3000 miles per hour (4800 kilometers per hour) or more.

"Coming up on SRB burnout," the first direct talk from Houston control.

Shuttle is now more than 25 miles (40.23 kilometers) up, and the solid rocket boosters have exhausted their fuel. At 27 nautical miles (50 kilometers), they will separate from the space shuttle, fire their own engines to move away from the speeding orbiter and external tank, and fall to the ocean to be picked up for reuse on another flight.

"Stand by for separation."

"Four . . . three . . . two . . . one . . ." Several muffled explosions indicate that the pyro-devices, explosive bolts, have fired, separating the SRBs from the shuttle. The acceleration is less now without the boosters. The space shuttle is proceeding toward space on its three main rocket engines, fed by the monstrous fuel capacity of the external tank.

The thrust builds up again as the shuttle becomes lighter. The engines throttle back to keep the thrust down to a 3-g level.

The curve of the earth is visible from the orbiter, and the sky has now turned completely black. Inside, the soft hiss of the cabin pressurization melts with the small vibration that is the only indication that the engines are still going. The crew is beginning to relax. It's almost over now. The fuel in the giant tank is nearly expended and main engine shutoff is coming up shortly.

The bottom drops out of the crew's stomachs as the engines shut down.

"Houston," says the command pilot. "We have shut down."

"Roger," comes from the ground, along with a string of figures—meaningless to everyone in the crew except the pilot and copilot. The space shuttle is now approaching 115 miles (185.5 kilometers) above the earth. It is a couple of hundred miles per hour short of orbital velocity. The crew waits for the separation of the external tank, which has nearly reached orbit with them. The explosive bolts fire and the tank spins away and falls back to earth in a ballistic trajectory. It will land somewhere in the ocean, far from human habitation, and will not be reused—a waste, certainly, but the tanks are cheap enough that recovery is not monetarily justifiable.

When fuel from the giant external tank is exhausted, the tank is separated from the orbiter. It will later enter the atmosphere and impact in remote areas of the ocean. The external tank is not designed for reuse, although it has been suggested that some tanks may be taken on up into low-earth orbit to form a base for construction platforms or habitability modules in orbit–space stations. (Courtesy Rockwell International Space Division)

Once the external tank has separated, the orbiter uses its orbital maneuvering system to perfect the orbit and for orbital rendezvous operations. (Courtesy NASA)

When orbital operations are completed, the orbiter fires its maneuvering system and deorbits. The speed is reduced, and the huge craft begins a descent through the earth's atmosphere. (Courtesy Rockwell International Space Division)

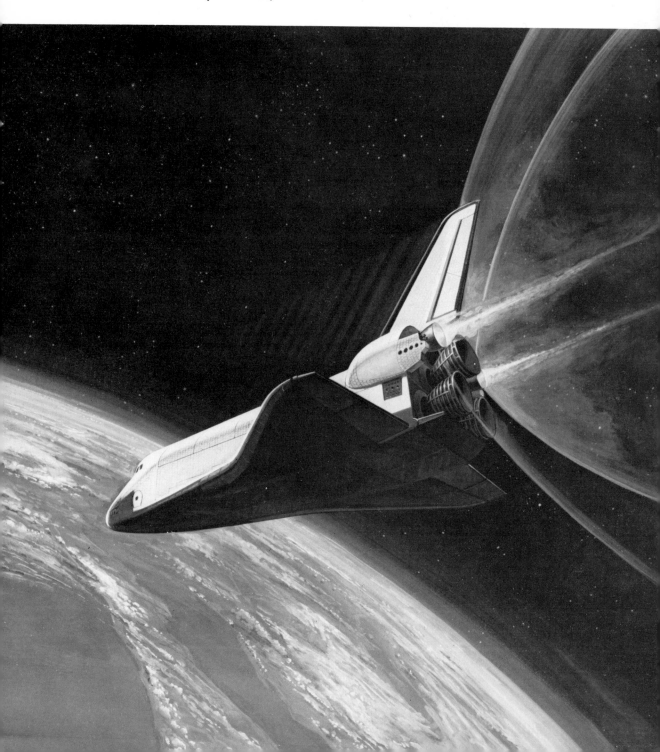

In the lower reaches of the atmosphere, the orbiter begins to fly like an aircraft. (Courtesy NASA)

"Coming up on OMS ignition," says the commander. The shuttle will use its orbital maneuvering system to gain the two or three hundred miles per hour it needs to achieve a stable low-earth orbit. The computers check with the ground and each other as the shuttle is pointed in the proper direction before the maneuvering engines fire.

Houston control asks for clarification of ignition. The pilot gives it as the very low thrust from the OMS comes on. Compared to launch, the thrust from the maneuvering engines is insignificant. A few minutes after ignition, the computer gives its answer. The orbital speed is perfect. Shuttle is "go" for orbit.

"We have OMS shutdown here," reports Houston.

"Roger."

The space shuttle is in orbit around the earth. Two pilots have both flown their fourteenth trips. One mission specialist just chalked up number five. The rest of the crew has had their first flight into space, three of them, essentially, as passengers.

The passengers unstrap slowly, floating about the mid-deck in the weightlessness of zero-g. It is something they will have to get used to. It will last six more days until the shuttle performs a deorbit burn and lands them on the space shuttle runway at Kennedy Space Center.

The week of work in space passes quickly for the passengers and the new mission specialist, slowly for the old hands. Deorbit burn comes up on Houston's mark exactly on time and on the button. Shuttle, now pointing in a direction opposite the orbital motion, fires the orbital maneuvering system. The OMS, acting as a retro-rocket, kicks the space shuttle out of orbit over the southwestern Pacific. The shuttle begins to slowly fall out of orbit as it drifts around the earth.

At reentry, the nose of shuttle is stuck up at a 40-degree angle. The earth is out of sight below. The entry into the earth's atmosphere is completely computer-controlled and so is the landing, although either of the pilots can take over manually if anything goes wrong. But nothing is quite as accurate as the computers, or as quick. Fifty times *each second* the computers monitor each other. They continually verify the health of the other computers, and if one becomes even momentarily erratic, the others gang up on it, pronounce it "sick," and cancel its decisions.

At 400,000 feet (121,920 meters), the first thin, wispy tendrils of the upper reaches of the earth's atmosphere brush the underside of the space shuttle orbiter. It is somewhere far above Hawaii. There is a slight flutter, a vibration in the orbiter structure, as it dips deeper into the heavier air. The temperatures of the outer skin, monitored very carefully aboard the ship, begin to rise from the air friction. At 300,000 feet (91,440 meters), the air rushing past the spacecraft's hull is a scream, and the surface temperature goes up to 2700 (1538.8 Celsius) degrees. Inside, it is normal. The heat shield composed of the ceramic bricks of the orbiter's outer skin is keeping the heat away from the orbiter compartments.

Near the end of a mission, the shuttle orbiter lines up with the runway for a landing. Shuttle can land at either Kennedy Space Center, Florida, or Vandenberg Air Force Base in California, when the latter becomes operational. (Courtesy Rockwell International Space Division)

Coming in. With landing gear down, the space shuttle goes down on the runway at nearly 300 miles per hour (500 kilometers per hour). (Courtesy NASA)

Opposite: Artist's rendering of shuttle night launch. (Courtesy NASA)

Shuttle lifts off from launch pad with main engines and solid rocket boosters firing. (Courtesy Rockwell International Space Division)

Opposite: With remote manipulator arm, shuttle can place a payload into orbit. (Courtesy NASA)

Opposite: Space shuttle orbiter comes in for a landing at Kennedy Space Center. (Courtesy Rockwell International Space Division)

Two views of shuttle glowing with the heat of re-entry. A 1952 painting of a "space shuttle" which Wernher von Braun proposed (*Across the Space Frontier*, courtesy Viking Press) is similar in design to the modern space shuttle. (Courtesy Rockwell International Space Division).

The life support system is working overtime but perfectly.

The hull temperatures drop at eight minutes after reentry. The orbiter is being steered like an airplane, or even more, like a giant glider. The reentry acceleration is only 1.5 g, which is low, but seems strange to the crew, especially the new members, after a week in orbit. At 160,000 feet, the speed is down from an orbital velocity of 17,600 miles per hour to 180 miles per minute and falling. Radio blackout has come and gone and the pilots are beginning an obscure chant with Houston control. Weather at the shuttle landing strip at the Cape is "CAVU": clear air, visibility unlimited.

At 120,000 feet, the orbiter executes a dog-leg turn to line itself up with Cape Canaveral. Gliding down, it drops like a brick—much faster than any normal airliner. The glide angle is more than seven times that experienced by a

Air brakes and wheel brakes on, the shuttle orbiter still needs plenty of room for landing. This photo was taken during the orbiter's second, free-flight test in the ALT series, October 12, 1977, at Dryden Flight Research Center, California. (Courtesy Jim Long Associates and NASA)

regular air traveler, except one going down for the last time. From the orbiter forward windows, mercifully blocked from the vision of the inexperienced in the crew, the view is nothing but earth rushing up at a fantastic speed.

Now below the speed of sound, still nosing toward the earth, the flight is released by Houston Control to Cape Approach. At 20,000 feet, the ship banks left and begins final approach.

"Altitude fifteen thousand," squawks Cape Approach. "Two minutes . . . two minutes. We have acquired your landing system."

"Roger, Cape."

The orbiter's computers are having a tête-à-tête with Cape Approach's computers in silent confrontations of symbols and numbers.

At 1700 feet, one-and-a-half miles from Kennedy Space Center's shuttle runway, the orbiter levels and drops down to 210 miles per hour.

Cape Approach is calling out altitudes: ". . . nine hundred . . . six hundred . . . four hundred . . . two hundred . . ."

At two hundred feet, the pilot's left hand hits a button on the cockpit instrument panel. "Gear down," he reports to Cape Approach. "Down and locked. Green lights," he repeats.

The shuttle hits on the runway with a bounce and the brakes send chatter through the landing gear.

The vertical rudder splits and becomes an air brake, slowing the shuttle quickly. Outside in the bright Florida sun the ground vehicles are screaming across the runway to meet the orbiter.

And out across the flat sand and marshes of the Cape, over the Indian River, people in Titusville are just sitting down to breakfast.

It looked so easy when you did it on paper—where valves never froze, gyros never drifted, and rocket motors did not blow up in your face.

Milton W. Rosen, rocket engineer, 1956.

2

THE BRICK MOON AND OTHER FANCIES

Riding an orbit 115 miles high, the space shuttle, upside down relative to the earth's surface, begins another mission.

"Houston, do you confirm our orbit?"

"Roger, you're stable and on the button within half a mile."

In the mid-deck, crew members who have never flown in space before drift weightless to the small round porthole in the shuttle orbiter's entry hatch. The black of space seen from the porthole is cut by a giant arc, the indistinct curve of the earth's atmosphere. Below the haze, the white clouds, and the dark patches of denser cloud is the multihued surface of earth–brown and greens of continents, some blue of ocean.

Looking through the porthole, they try to see Cape Canaveral, already obscured by the clouds which brewed up in the Caribbean and swarmed over the west coast of Florida not long after launch.

Apollo was the result of a magnificent dream—a landing on another world. It was a dream which began, at least as far as ancient literature shows, with the writing of *Vera Historia* by a Greek author named Lukian. *True History,* written about

A.D. 180, describes a voyage to the moon by means of a giant whirlwind which lifts the heroes' ships from the ocean to the lunar surface.

The idea was kept alive by Johann Kepler's *Sleep,* a guidebook for emigrants to the moon written in 1634; *The Man in the Moone: A Discourse of a Voyage Thither* by Bishop Francis Godwin, 1638; the novels of Jules Verne, Alexandre Dumas, and an anonymous Englishman who wrote *The History of a Voyage to the Moon* around 1865; and H.G. Wells's *The First Men in the Moon.* Innumerable science fiction novels and short stories written in the twentieth century have dealt with the dream of landing on the moon, so no one involved in the dream was taken much by surprise when it actually happened at last.

Space shuttle is the result of a different dream, one which is much more recent than the literature of the Greeks or the gentle fumblings of the seventeenth century. It is the final realization of living and working in the space around the earth, of building space stations, space structures, and ornaments of commerce high above the atmosphere where people spend days and weeks, even months and perhaps years, at tasks which benefit those below and prepare for the eventual exploration of that above.

It is the final product of the dream of a vehicle which could supply materials to space on a near-daily basis, which would not be expendable and monstrously expensive, and which could be used to open space for all mankind. It is the first step toward the first real, manned space station with regular traffic between space and the earth below.

The idea of a space station, or the use of near-space with regular supply rockets, is quite recent. It first appears in 1870–71 in a story serialized in the *Atlantic Monthly* called ''The Brick Moon.'' The author was Edward Everett Hale. It was released in book form by Little, Brown in 1899. The story and the book proposed the launching of a giant, artificial satellite made entirely of brick. It was to be placed in a polar orbit at an altitude of 4000 miles (6400 kilometers) to aid navigation. The method of propulsion for the satellite was novel, if nothing else. The brick moon was to be shot into space by rolling against a rapidly spinning flywheel.

Unfortunately, that brick moon was inadvertently launched early, complete with thirty-seven construction workers and some of their families. Having no way down, the workers set up house on the brick moon. Plans were made to supply them from earth.

Another early discussion of a space station, and a shuttle rocket to supply it, is Kurd Lasswitz's *Auf Zwei Planeten,* published in 1897. In Lasswitz's book, the establishment of a space station is the key to future space travel.

Since the idea of a permanent space station appeared so late in literature, writings about shuttle-type vehicles and space stations quickly became much less fanciful than some of the science fiction of lunar and planetary travel. Authors stayed much closer to the work being done by rocket pioneers in this country and

One of the fathers of space travel, Konstantin Eduardovich Tsiolkovsky, featured on the cover of a Russian spaceflight book published in Moscow, 1954. (Author's collection)

Dr. Robert H. Goddard in his laboratory in New Mexico, 1932. The rocket shown is probably one of his efforts, which rose to an altitude of 7500 feet (2286 meters). (Courtesy Science Service)

others, especially Germany. The last of the very early ideas of a permanent space station—and the rocket necessary to service it—was contained in a book by the Russian scientist Konstantin Eduardovich Tsiolkovsky, called *Dreams of Earth and Sky,* published in 1895. In this book, the Russian space pioneer envisioned an artificial space satellite orbiting at a height of 200 miles (320 kilometers).

Following the publication of *The Brick Moon, Auf Zwei Planeten,* and *Dreams of Earth and Sky,* all slightly before the turn of the century, there is a gap of about twenty-five years before the subject is taken up again. During that interval, the means to accomplish all of the dreams in the books was invented: the liquid-fuel rocket.

There are three fathers of space travel, and one is the above-mentioned Tsiolkovsky. His first published article, ''Free Space,'' written in 1883, had accurately described the condition of weightlessness in space. By 1898, he had worked out the basic theory of rocket propulsion. He had also established figures showing the amount of propellant needed for a rocket of a given mass. In his article ''Exploration of Space by Reactive Devices'' (1903) he not only demonstrated that space travel by rocket was possible, but also designed a rocket that would work on liquid hydrogen and liquid oxygen, the same fuels used by space shuttle. Tsiolkovsky also anticipated the step rocket of multiple stages, such as the Saturn V used to launch the Apollo missions. Even space shuttle is a step rocket, the solid rocket boosters being one step.

Tsiolkovsky was, however, always a theoretician. The second father of modern space travel was an American, Robert Hutchings Goddard. His first rocket patents were taken out in 1914. He designed and built the world's first liquid-fueled rocket, which flew on March 16, 1926, using liquid oxygen and gasoline. It stayed aloft for only 2.5 seconds and rose to an altitude of 41 feet (12.5 meters). It was Goddard who ''got his hands dirty'' with the innumerable details of rocket construction and behavior. When he was finished with his experiments and launchings, he held more than two hundred patents, forming the basis for nearly all rocket development in America throughout the space program. One of his largest rockets was 22 feet (6.6 meters) in length, 18 inches (45 centimeters) in diameter, and weighed 450 pounds (200 kilograms). Robert Goddard was nearly alone in this country in the development of the rocket, and his work was unrecognized for most of his life, at least as far as the general public was concerned. But Wernher von Braun, whose V-2 rockets during World War II made the rocket a terrible reality, once remarked, in 1947, that ''Dr. Goddard was ahead of us all.''

The third father of space travel was Hermann Julius Oberth. Like Tsiolkovsky, he was primarily a theoretician, but his writings were a major inspiration to the budding German interest in rocket research, which ultimately solved many of the problems of large rockets and paved the way for the Russian and American space programs after World War II. Oberth drew designs for a liquid-fuel rocket as early as 1917. His doctoral thesis was rejected in 1922, along with his designs of a rocket, but fortunately he turned the manuscript into a book called *Die Rakete zu den Planetenräumen (The Rocket into Interplanetary Space)*.

Published in 1923, this book broke the twenty-five-year gap in literature on space stations and rockets to construct them. In *The Rocket into Interplanetary Space,* Oberth proposed orbiting space stations. On page 86 discussion begins about rockets carrying two men. ''If we let such rockets of the largest size (the two-man) move around the earth in a circle, they will behave like a small moon.

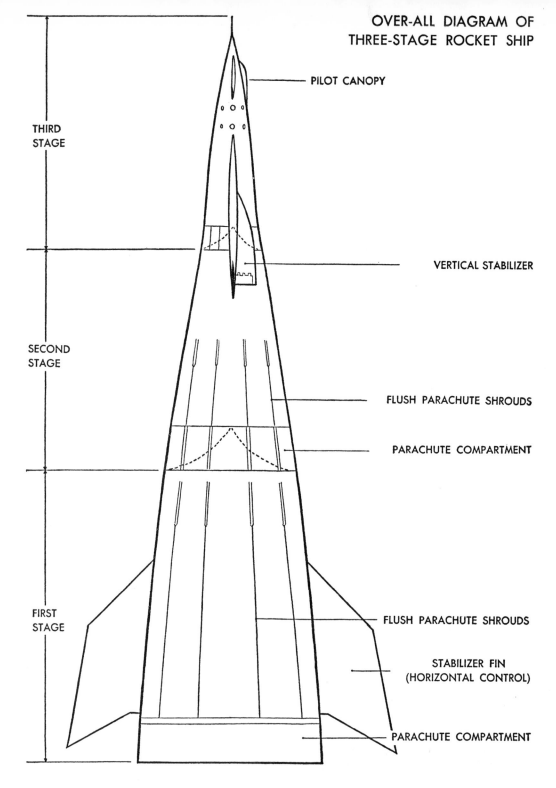

OVER-ALL DIAGRAM OF THREE-STAGE ROCKET SHIP

THIRD STAGE

SECOND STAGE

FIRST STAGE

PILOT CANOPY

VERTICAL STABILIZER

FLUSH PARACHUTE SHROUDS

PARACHUTE COMPARTMENT

FLUSH PARACHUTE SHROUDS

STABILIZER FIN (HORIZONTAL CONTROL)

PARACHUTE COMPARTMENT

*In **Across the Space Frontier**, published in 1952, Wernher von Braun proposed a three-stage rocket "shuttle" which would be used to build and supply a huge space station. The third stage of the rocket is basically similar to the space shuttle orbiter. (Courtesy Viking Press)*

Such rockets no longer need to be designed for landing. Contact between them and the earth can be maintained by means of smaller rockets so that the large ones (let's call them observing stations) can be rebuilt in the orbit the better to suit their purpose.''

This 1923 concept was the forerunner of the *Skylab* and space shuttle: the main ship stays in space, not being designed for landing (a space station); supplies are brought by reuseable transport rockets of smaller size; and the gradual build-up of space structures is similar to the way NASA is planning for the 1980s and beyond.

All proposals for space stations required some sort of transport rocket to ferry equipment up to the station orbit. It was from the concepts of these stations that the reuseable space shuttle was born: a rocket which could take off from earth, ferry freight up to the space station or other orbiting structure, and return to earth for another load.

In 1929 an Austrian named Potočnic, writing under a pen name, proposed a complicated space station which, of course, required the use of shuttle rockets.

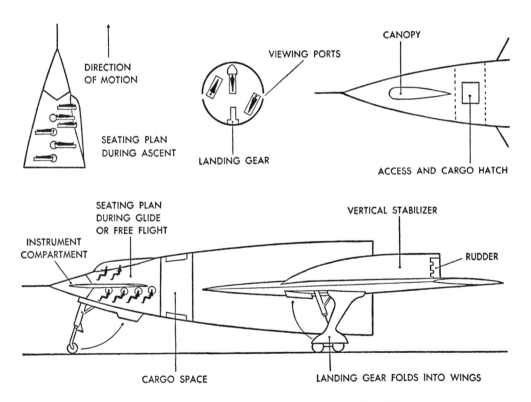

Von Braun's third stage had a capacity for six crew members and would land like an airplane on a runway at a space port either on the coast of Florida or on a remote island. (Courtesy Viking Press)

Perhaps its most interesting characteristic was its shape. Hermann Noordung's station was a giant wheel which he called a *Wohnrad* or "living wheel," and was 100 feet (30.48 meters) in diameter. The shape, if not Noordung's rather clumsy ideas, survived in countless science fiction stories and in the movie *2001*.

The first transport rocket containing the ideas which later became part of the space shuttle was designed by Eugen Sänger, an Austrian pioneer of astronautics. He developed the concept of a reuseable space transporter in the late 1920s. By 1933 he was designing a rocket plane. During the height of World War II, he concentrated on the idea of a long-range winged bomber, boosted by a rocket sled and its own engines. This work led to the post-war studies on reuseable rocket systems, of which space shuttle is the end product. Just before his death, Sänger was drawing designs for two-stage winged space transporters which could be used to ferry materials and men up to orbiting space structures—the exact role of the space shuttle which would become operational nearly fifteen years after his death.

Sänger was not the only one interested in space transport vehicles. Other designs were worked on during World War II at the German rocket base at Peenemünde by Wernher von Braun and his rocket design group. After the war, von Braun formalized his ideas in a classic called *Across the Space Frontier* (1952), edited by Cornelius Ryan. Other authors in the book were the famous science writer and popularizer Willy Ley and astronomer Fred L. Whipple. The

*"Landing of the third stage," from **Across the Space Frontier**. (Courtesy Viking Press)*

illustrator was Chesley Bonestell, one of the best-known space illustrators then or now. The key to the concepts in the book, which was developed from a series of scientific articles called ''Man Will Conquer Space Soon'' in the now dead *Collier's,* was a giant step rocket of three stages, which would service a wheel-shaped space station orbiting at an altitude of 1075 miles (1729 kilometers).

The rocket, described by von Braun in a chapter ''Prelude to Space,'' was 265 feet (80.56 meters) high and 65 feet (19.76 meters) in diameter. Taller than a twenty-four-story office building, it weighed as much as a light cruiser. The first stage had fifty-one rocket motors, while the second stage had thirty-four. But it was the third and final stage of von Braun's supply rocket which so resembled, at least in concept, the present-day space shuttle.

The third stage, equipped with five rocket motors, could carry a space cargo of 36 tons—very close to that which the space shuttle of the 1980s will carry. In addition, and more importantly, von Braun's rocket was designed to take off, like a rocket, from a base on the east coast of America (''. . . the Air Force Proving Ground at Cocoa, Florida,'' was an example of a suitable site), fly to orbit, unload its men and supplies, and then glide on wings down to an earth landing at a normal airstrip, just like a large airplane.

Dr. von Braun's ''Proving Ground'' at Cocoa, Florida, is, in fact, what is now Cape Canaveral, Kennedy Space Center. His idea of the winged third stage which lands like an airplane is the same used with the space shuttle orbiter, the first of which, Orbiter 101, is called *Enterprise* after the ''Star Trek'' spaceship.

The 1952 concept even anticipated the approximate size of the crew of a shuttle rocket: von Braun's proposed six—two pilots and four crewmen. Down to small details, the book anticipates the scene which would be enacted at Cape Canaveral more than eighteen years later: ''At the launching area, the heavy rocket ship is assembled on a great platform. Then the platform is wheeled into place over a tunnel-like 'jet deflector' which drains off the fiery gases of the first stage's rocket motors. . . .'' Not a poor description of the crawler at the Cape, nor of the scene at Launch Complex 39, pad A, in 1979 when the very first, real, space-shuttle flight takes place.

The plan also anticipated salvage ships rescuing the spent first stages of the rocket for reuse on a later flight, in the same way the solid rocket boosters (SRBs) of the space shuttle are salvaged from the ocean drop zone and reused on a later flight.

In addition to the theoretical designs, which were a result of World War II rocket research, the practical aspects were developed, too. The V-2 was the world's first real space rocket. After the war, one went up almost 114 miles, but it was never designed as a research vehicle. Following the V-2 was a series of rockets developed in the United States from the Goddard patents, the V-2 designs, experiments with our own Viking rockets of 1949 to 1955, and others. The

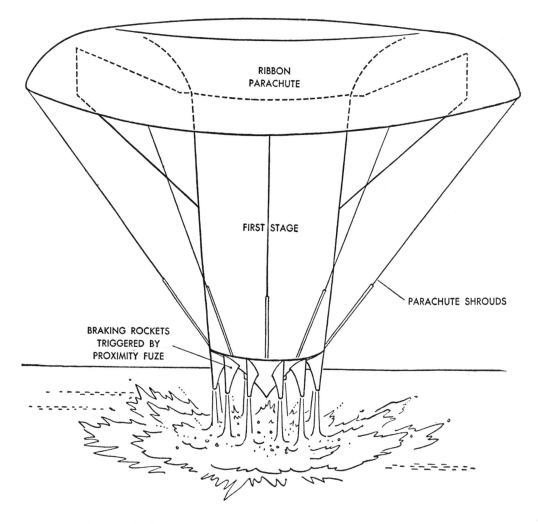

RIBBON
PARACHUTE

FIRST STAGE

PARACHUTE SHROUDS

BRAKING ROCKETS
TRIGGERED BY
PROXIMITY FUZE

As with space shuttle, the ''main stage'' in von Braun's shuttle rocket would be picked up in the ocean for reuse. The space shuttle's ''main stage'' is the SRBs. (Courtesy Viking Press)

culmination of all this rocket research over the years 1955 to 1979 were the basic propulsion devices of the American space program: Redstone, Jupiter-C, Atlas, Thor, Atlas-Agena, Titan, Titan III, Saturn V, and Titan Centaur. And the space shuttle.

NASA officially began investigating the space shuttle designs in early 1968 and by 1972 had fixed the general configuration. In early proposals, the shuttle was a winged, reuseable booster as well as a winged, reuseable orbiter. Not surprisingly, there are innumerable comparisons between the current design for

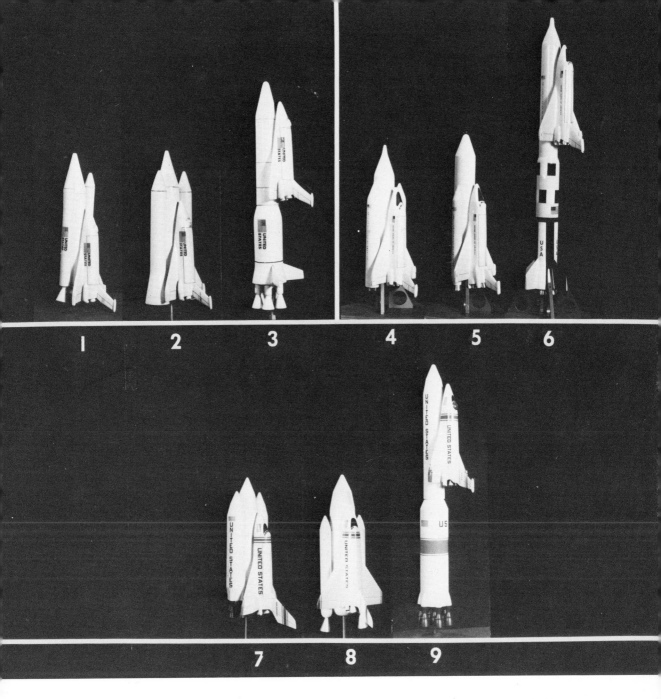

Various concepts of the Space Transportation System in 1972. (1) a recoverable liquid-fuel booster (2) an expendable solid-fuel booster (3) a different concept of a recoverable, liquid-fuel rocket booster (4) an expendable solid rocket booster (5) a recoverable liquid-fuel rocket booster (6) a recoverable booster and a manned reuseable orbiter (7) a recoverable liquid-fuel booster (8) another expendable solid rocket booster and (9) the most unusual and expensive of them all, a recoverable unmanned booster and manned orbiter. 1, 2, 3 were designs of McDonnell-Douglas Corp./Martin Marietta; 4, 5, 6 were North American Rockwell/General Dynamics; 7, 8, 9 were Grumman Aerospace Corp./Boeing Co. (Courtesy NASA)

space shuttle and the older 1952 von Braun step rocket. Shuttle is not, however, nearly as large. The von Braun rocket weighed 14,000,000 pounds (6,350,400 kilograms), while the shuttle weighs only 4,400,000 pounds (2,000,000 kilograms). Shuttle's orbiter, equivalent to von Braun's original third stage of the step rocket, is about the size of a modern jet liner, 122 feet (37.2 meters) long, with a wingspan of 78 feet (23.8 meters).

The space shuttle first flew on February 18, 1977. It was held captive and unmanned on a 747 aircraft for the test. Following this test were four other captive, unmanned tests, ending in March 1977. Manned captive flights ceased on July 26, 1977, with flight number three. Following that, there were manned free flights, complete with landings, up to October 26, 1977, using the long concrete runway of Edwards Air Force Base, California. All of the flight testing and landings indicated that the shuttle orbiter was ready for space, after some additional structure testing. The first manned orbital flight of the space shuttle from Cape Canaveral was scheduled for September 1979, less than one hundred years after the publication of *The Brick Moon,* slightly more than fifty years after Goddard experimented with the world's first liquid fuel rocket, and less than three decades after von Braun's step rocket and winged reuseable third stage appeared in *Across the Space Frontier.* The first *operational* flight, Space Shuttle Flight 7 (SS-7), is scheduled for February 1981.

It is not really necessary to look too far into the future; we see enough already to be certain it will be magnificent. Only let us hurry and open the roads.

Wilbur Wright

3

APPOINTMENT IN ORBIT

Shortly into the first orbit, the shuttle's giant payload doors swing open, exposing the coolant radiators to space. The cargo bay, pointed toward the earth below, is illuminated by the bright earthshine of white clouds and sunlight reflected from the land and ocean masses. The coolant radiators remove excess heat from the orbiter made by the electrical systems and the life-support equipment.

From the aft flight deck windows, looking across the cavern of the cargo bay, the crew can see the giant tail and rudder of the space shuttle and the bulging pods of the maneuvering engines.

Farther around in its orbit, the shuttle gets its first sunset, a sunset never seen on earth. The glaring disk of the sun, blazing away without an atmosphere to absorb and mute the light, approaches the curve of the earth. As the fiery circle touches the earth's atmosphere, the shuttle drops into shadow, the night of eclipse. The sun moves further into the layers of the earth's atmosphere, and colors spread away on each side. The dense air contorts the light to reds and yellows, blues and greens, a giant prism. The sun disappears below the rim of the earth and the colors fade.

Not even in the high desert does the sunset pass so quickly. The sun is gone before the mind can fully accept the coming darkness.

But the darkness comes, black on black velvet with points of brilliance woven in. A carpet of stars, a hundred constellations, spread across the sky.

The shuttle is scheduled to keep its first appointment in orbit in September 1979, with orbital flight test missions. After six test flights, the space shuttle will become operational, probably in February 1981, although earlier predictions were for operational basis in May 1980. While the early orbital flight tests will carry only a crew of two—flights 5, 6, and 7 will carry four—the operational flights will have a full complement, depending on the purpose of the mission. Space shuttle can carry six to ten people on each flight, lasting from seven to thirty days, at altitudes ranging from 115 to 690 miles (185 to 1110 kilometers). NASA is currently investigating the possible impact of certain 1979 funding for shuttle. Therefore, it is possible that low funds for the project could cause the first orbital flight test to come as late as March 1981, with a corresponding delay in the operational status.

Shuttle is a relatively complex vehicle, composed of the orbiter, which is a winged spacecraft about the size of a medium jet liner; the external tank, a gigantic cylinder for liquid oxygen and liquid hydrogen, feeding the orbiter engines; and the solid rocket boosters (SRBs) which provide part of the huge initial power to lift shuttle from the pad at Cape Canaveral. The orbiter and the solid rocket boosters are reused after each flight—the boosters are recovered by ships—while the external tank is targeted to land in the sea and never recovered.

Original designs for the space shuttle from 1968 to 1972 called for a fully reuseable system, in which the fuel was carried in a winged and piloted booster which would also land, like the orbiter, back at the Cape. For various reasons, including cost, this system was discarded in favor of the expendable external tank. Also, the later space shuttle looks considerably more trim than some of the former designs.

Various names for the space shuttle, both official and unofficial, have been associated with the program. The entire package of orbiter, ET, and SRBs is called the Space Shuttle Transportation System (SSTS) although the term "space shuttle," or simply "shuttle," has been applied to the whole package. Technically, it is only the orbiter which actually goes into space, often called the shuttle orbiter. And, of course, OV-101 (the manufacturer's designation for Orbiter Vehicle-101) has been named *Enterprise*. Three other shuttles will be named *Discovery, Challenger,* and *Atlantis.*

Unofficially, the orbiter has been nicknamed the Flying Brick, mainly because of the thermal protection system of ceramic tiles on the orbiter outer skin,

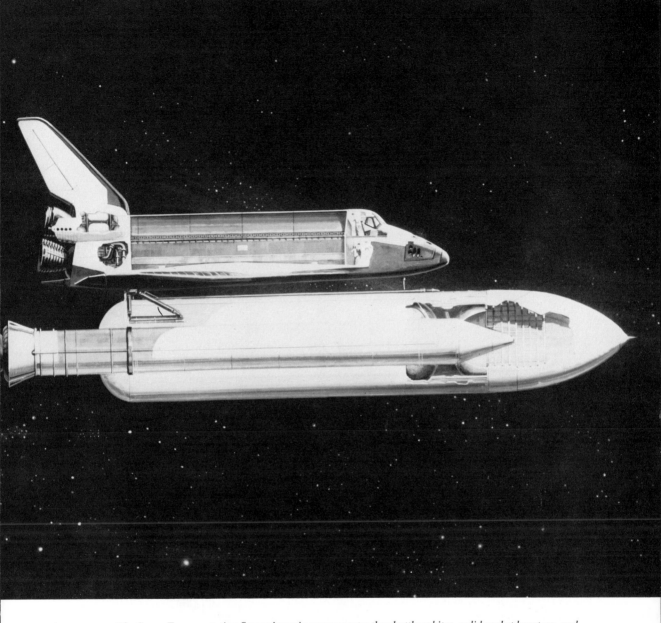

The Space Transportation System's main components: the shuttle orbiter, solid rocket boosters, and giant external tank in cutaway. (Courtesy NASA)

which protects it from the heat of reentry. It has also been called the Gooney Bird of Space, a reference to the famous DC-3 aircraft which did so much to expand and promote commercial aviation over a period of thirty or more years. It has been called ugly, but no one ever said spaceships have to be pretty, and there are many who disagree with that value judgment, anyway.

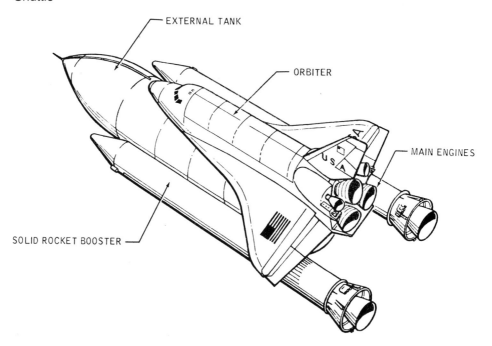

EXTERNAL TANK

ORBITER

MAIN ENGINES

SOLID ROCKET BOOSTER

The orbiter, with its payload cargo bay and crew module, will be dealt with in a later chapter. To get the crew and cargo into space in the first place, it is the other parts of space shuttle which are most important: the SRBs, the ET, the main rocket engines, the orbital maneuvering system, and the reaction control system. To get the crew down again is a job for the command pilot and the protection system against the extreme heat of reentry.

The solid rocket boosters, known by the initials SRBs, are attached to the side of the external tank. They are a solid propellant rocket—a giant skyrocket—and are only ignited *after* the space shuttle main engines have been ignited and are verified as providing thrust. It is the solid rocket boosters, together with the main engines of the shuttle, which lift the ship off the launch pad.

The "gunpowder" in the solid rocket boosters is a mixture of aluminum perchlorate powder, which acts as an oxidizer, and aluminum powder, which acts as a fuel. There is also an iron-oxide powder for a catalyst and a polymer to bind the mixture together. The polymer also acts as a fuel.

When the solid rocket boosters have done their job of powering the space shuttle, along with the main engines, to an altitude of about 31 miles (49.94 kilometers), they are separated from the space shuttle by pyrotechnic devices—explosive bolts or a similar arrangement. Eight solid-propellant separation motors, four forward and four aft on the solid rocket booster, push the boosters away from the space shuttle and the external tank.

The nose cone of the solid rocket boosters contains electronics, a safety-destruct system in case of accident, the four forward separation rockets, a drogue chute, and recovery gear.

After separation from the speeding shuttle, the solid rocket boosters descend to a predetermined altitude, and the nose cone fairing is jettisoned. The drogue chute comes out, slowing the solid rocket booster initially, then a main parachute pops up, and it floats down to a relatively soft, saltwater landing in the Atlantic somewhere east of Cape Canaveral. A recovery crew with a special barge recovers the solid rocket boosters and tows them back to the Cape, where they will be examined, maintenance will be performed, and the fuel and oxidizer replenished. After a few days, they will be ready for another shuttle flight to space.

It is difficult to describe the external tank, the ET, except by figures which cannot convey the size of the thing. It is 154.2 feet (47 meters) long and 27.5 feet (8.38 meters) in diameter, about the size of the body of a 747. It completely

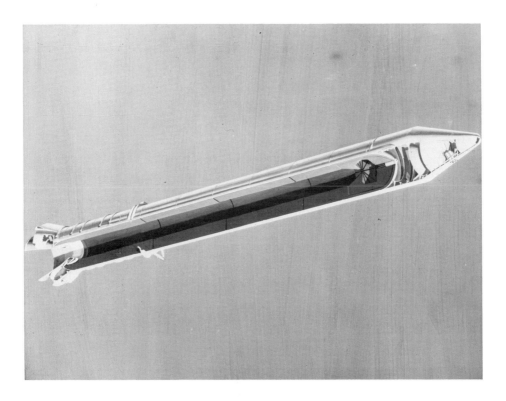

The solid rocket booster (SRB) of the space shuttle. Forward and aft maneuvering engines help separate the SRBs from the orbiter. Electronic gear and parachute equipment in the nose of the SRB allow it to land in the ocean and be picked up by ships. It can be refurbished for use on a later shuttle flight. (Courtesy NASA)

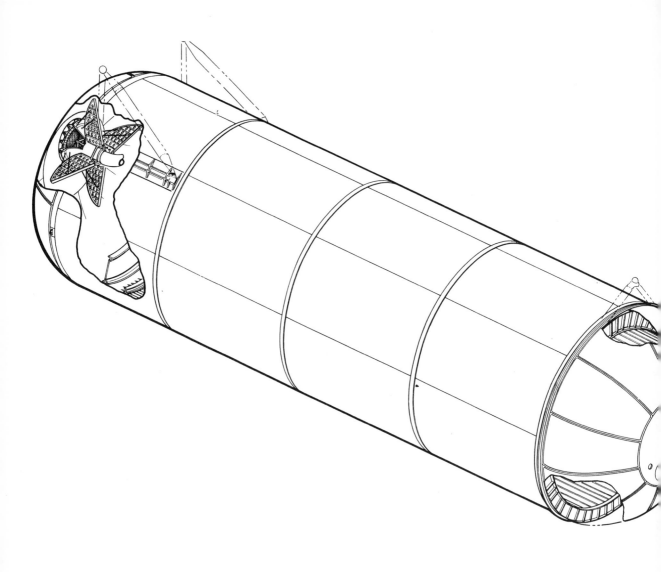

The structure of the external tank: liquid hydrogen tank, inter-tank, and liquid oxygen tank. (Courtesy Martin Marietta Aerospace Michoud Operations)

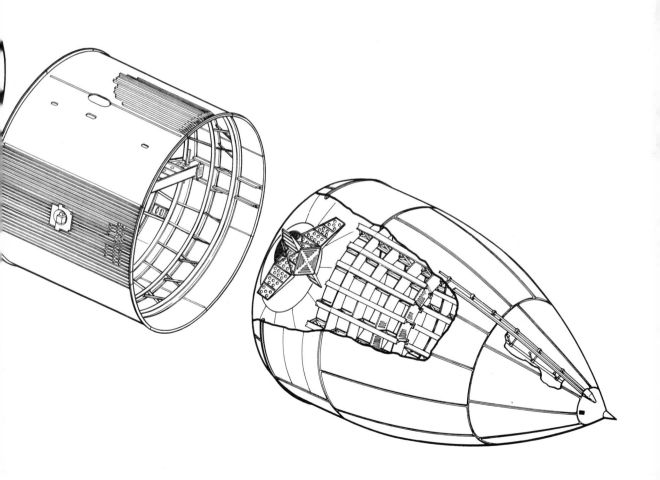

dwarfs the shuttle orbiter and nearly everything else around it, except the giant buildings at the Cape.

The external tank, a much more complicated device than it might at first seem, is made up of two tanks—a liquid oxygen tank and a liquid hydrogen tank—and an intertank connector. The nose piece is the liquid oxygen tank which has a modified cone shape, known as an ogive shape, to reduce the aerodynamic drag and generated heat. The intertank is the load distributor for the structure, and it joins the two fuel tanks within the external tank. Instrumentation and access to the launch pad are incorporated into the intertank.

Of all the space shuttle components, the ET, though appearing quite simple, was perhaps one of the hardest to manufacture. It was so large that special tools had to be designed and constructed, and special equipment had to be manufactured, to make the first external tank. One tool, which rolls the aluminum panels into a circle that forms the structure of the tank, was nicknamed the Cigarette Roller because of the way it rolled the flat sheets of aluminum into the cylinder of the tank. The machine, known more properly to the manufacturer as the Baltimore Tool, is 45 feet wide, 31 feet high, and 48 feet long (13.68 x 9.42 x 14.59 meters).

For the first truly mass produced piece of space hardware—the external tank—364 special tools were made, including 34 which are large, permanent major fixtures. Since the external tanks are not reused in a space shuttle mission, plans call for the manufacture of perhaps forty-five to sixty tanks per year in the late 1980s to accommodate the total number of planned flights of space shuttle through the 1980s, expected to be between five hundred and six hundred flights.

The external tank separates from the space shuttle just prior to insertion into the orbit selected, usually at an altitude of about 69 miles (127.65 kilometers). It follows a ballistic entry path—falls in a long curve—and impacts in the ocean away from inhabited areas. It is not recovered or reused.

The huge supply of liquid oxygen and liquid hydrogen fuel in the external tank is fed into the space shuttle main engines, or SSMEs. There are three shuttle engines in the aft section of the orbiter. They are fed by pressure in the external tank, generated by gaseous helium from the launch pad, through 17-inch (43.18-centimeter) ducts. From these, main supply fuel lines feed three smaller lines to the main engines. The smaller lines to the main engines are 12 inches (30 centimeters) in diameter! Once the SSMEs have started, the pressure for fuel feed is generated within the shuttle system.

The space shuttle main engines are a thirsty lot, consuming about 1122 pounds of propellant per second (508 kilograms). The thrust is enormous, as it must be to lift the giant space shuttle up to earth orbit. Each of the three engines has a throttle and is designed to last for more than fifty-five missions, which is equivalent to seven and a half hours of operation.

This is the forward section of the external tank, in the manufacturing facility at Michoud, Louisiana. The tank is 27.5 feet (8.38 meters) in diameter and 154.2 feet (47 meters) in length. Note the size of the workmen on the rim of the tank. (Courtesy NASA)

As in aircraft maintenance, the space shuttle main engines have a flight record. This gives the engine operating history, helps keep track of repair schedules, and extends the total engine life.

The external tank, the space shuttle main engines, and the solid rocket boosters are designed to take the space shuttle up to orbital speed. From that point, the flight of the shuttle is shifted to the orbital maneuvering system. Through two engines in pods on the tail of the shuttle orbiter, the orbital manuevering system provides the thrust to make the orbit insertion, follow an orbit, make a transfer from one orbit to another, and deorbit to begin the descent to earth, landing on the shuttle strip at Cape Canaveral.

In addition to the orbital maneuvering system in the two pods, there is the reaction control system, or the RCS. The RCS on the orbiter provides the power for small velocity changes along the orbiter axis and also controls the altitude control (pitch, yaw, and roll) at altitudes over 70,000 feet (21,336 meters).

The two orbital maneuvering system engines are reuseable for more than one hundred shuttle missions, can start and restart more than one thousand times, and can run up more than fifteen hours of cumulative firings. Each of the orbital maneuvering pods has a helium tank for pressure, an oxidizer, and a fuel tank. Propellants, in the case of the orbiter maneuvering system, are a combination of nitrogen tetroxide as an oxidizer and monomethyl hydrazine as a fuel. The components of this fuel are hypergolic, which means they ignite on contact.

For extended space shuttle missions, and depending on the weight of the cargo in the shuttle payload bay, extra fuel tanks, or propellent kits, can be added to the cargo bay, so the shuttle can maneuver considerably while in earth orbit, if necessary.

The reaction control system is located in three places on the orbiter: in the left and right pods of the maneuvering system on the aft fuselage, and one in the forward fuselage area.

Each of these three reaction control system areas has two helium tanks for exerting pressure on the fuel and oxidizer tanks. The reaction control system uses the same type of fuel as the orbital maneuvering system. For perfect control during a flight, it includes twelve small primary engines and two vernier engines. The forward reaction control system has fourteen small primary engines and two vernier engines. The primary engines are small but are the real control factor of the space shuttle. They can be used on more than one hundred missions, and each of them can stand fifty thousand starts, more than twenty thousand seconds of cumulative firing. The vernier engines make very small adjustments or correc-

Flakes of frost shake loose from a test main engine, one of three like those used on the shuttle orbiter. This 1.6 second firing of the engine was at Bay St. Louis, Mississippi, at NASA's National Space Technology Laboratories. (Courtesy NASA)

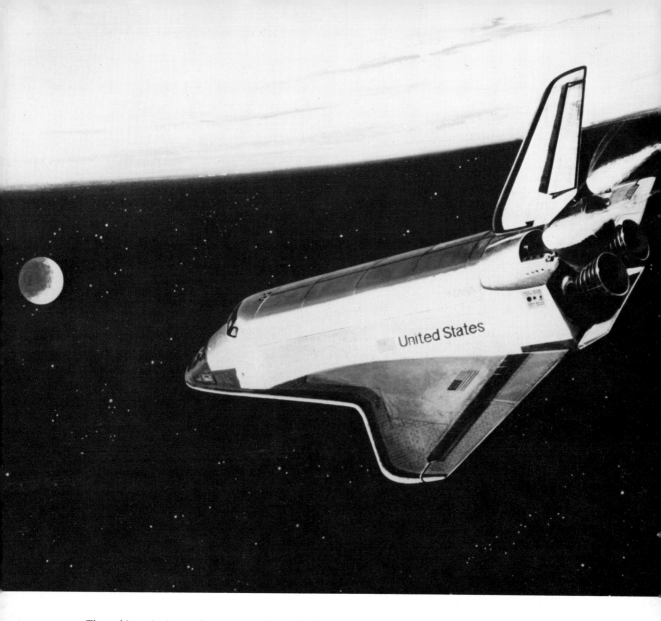

The orbiter, in inverted pose, uses the orbital maneuvering system to move about in space. (Courtesy NASA)

tions to the shuttle's position in space. They are similar to the primary engines, but are capable of *500,000* starts and over 125,000 seconds of firing.

Fuel from the orbital maneuvering system can be used in the reaction control system through an interconnect. The interconnections provide the means to jockey fuel back and forth from the left and right maneuvering pods, as well as the ability to use the main fuel supply of the pods, and the payload area extra-fuel kits, when they are fitted.

For landing, the space shuttle simply uses the orbital maneuvering system to deorbit, and then flies down to a landing on its landing gear. During the reentry phase, the orbiter is protected by a thermal protection system, because of the heat generated as the shuttle strikes the upper reaches of the atmosphere at high speeds. Temperatures on an unprotected surface on reentry can exceed 350 degrees (176 degrees Celsius), and some areas of the orbiter external skin can be greater than 2300 degrees (1260 degrees Celsius).

The thermal coating, which protects the orbiter in the most critical areas, is a form of ceramic brick. This was mentioned earlier as the source for the orbiter

Bricks of pure silica are the thermal protection for the orbiter during the heat of re-entry. Nearly 34,000 tiles, in various sizes, cover the upper and lower surfaces of each orbiter. These unique tiles have led to the nickname, "The Flying Brick." (Courtesy NASA)

nickname, The Flying Brick. Other types of thermal protection for the orbiter are used in less critical areas.

For some people, the most important part of the Space Shuttle Transportation System is the orbiter crew module and the payload cargo bay. That is, after all, what the shuttle is all about: the men and women who will go into space and the materials and devices which they will take there. It is a fascinating and complicated space ship, not too unlike, as was pointed out earlier, some of the great dreams of the 1950s.

There are flying grandfathers. But I intend to be an orbiting *grandfather.*
Wernher von Braun, shortly after the *Apollo 11* lunar landing, July 20, 1969

4

ABOARD AN ORBITER

In the cramped galley of the space shuttle, the crew is having dinner in space: shrimp cocktail with sauce, beefsteak, broccoli au gratin, and cocoa. The earth below, faintly illuminated by moonlight, is a week away at mission's end. Off in the other direction, the moon, two days after first quarter, pours soft light into the windows of the flight deck. It is the same moon seen from Apollo, looking as far away as it ever does from the earth's surface. The shuttle is only 115 miles closer than earth to the dead powder surface of our only companion satellite. A million years have passed, and a million million and there has been no change on the moon. It lies there always, dead and yet alive in dreams, poems, and songs.

A million years before, there was no mankind that this crew of a space shuttle would recognize. Even ten thousand years backward in history would not find a man writing his own name, seeking language, conveying information by curious symbols which might be vaguely recognizable later in the constant flow of symbols across the screens in the cockpit of the spaceship hovering over the earth. If they think of it at all that late night in the Moon Room while dining, they might laugh nervously or smile in melancholy at the presumptions of man, the perverse drive which raised great monoliths at Stonehenge, peered and probed into this and that function of nature, grabbed the kernal of all creation away from the atom,

61

and pushed the frail body known as a human being out into the great deeps beyond the atmosphere of the planet of birth.

They are tired, these space travelers. It is night time on the earth below and night time for them. While they dine between the earth and the moon, the shuttle continues circling.

Wernher von Braun did not live to be the spry and healthy seventy-year-old who would be a passenger on the space shuttle he had helped design. But if he had lived, he *could* have been one. He was an excellent pilot in good physical condition, and the space shuttle is gentle on its crew. He could have qualified as a mission specialist for NASA or a payload specialist for a dozen private companies.

The heart of the Space Transportation System is the shuttle orbiter. It is a double delta-winged craft, about the same physical size as a DC-9 jet airplane. Its huge cargo bay can hold a payload the size of a Greyhound bus. A normal crew is composed of two astronauts; a NASA mission specialist or two; and one to three payload specialists depending on what the space shuttle is carrying into orbit that day.

In an emergency, orbiter can carry ten people. On a rescue mission, three crew men or women would be sent up in an orbiter to take on the crew of seven of a crippled mission.

Originally there was to be an orbiter fleet of five, consisting of two vehicles bought in the developmental phase of the space shuttle and three production orbiters. The developmental orbiters were to be refurbished later for use as operational vehicles. Funding cuts and general space-program budget problems have modified the initial plan. Orbiter Vehicle 101 was completed, named the *Enterprise,* and put through the extensive testing program known as the approach and landing test (ALT). It also was used, mated to a test external tank and other shuttle hardware, as a test vehicle for the entire shuttle system. When the tests are finished, it will be returned to the manufacturing facility to be refurbished for operational use in the 1981–82 period.

The *Columbia* was completed and will be the first orbiter to be launched into space during the orbital test flights (OTLs) beginning in 1980. Four of these tests will feature landings at Edwards Air Force Base in California, while the last two flights of the six planned will end at the Cape Canaveral shuttle-landing strip.

According to John F. Yardley, associate administrator for Space Transportation Systems of NASA, speaking before the Subcommittee on Space Science and Applications, Committee on Science and Technology, United States House of Representatives, on September 25, 1978, ''Long lead work has begun on the

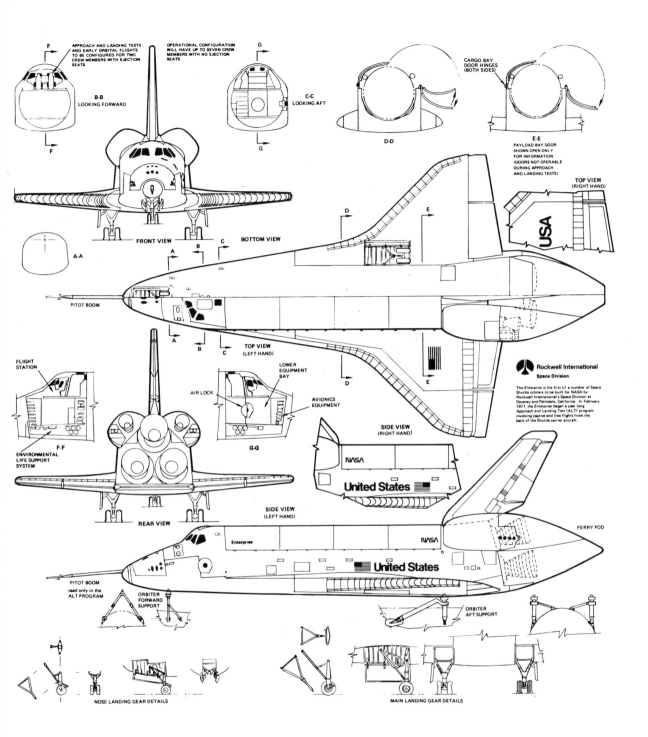

An artist's concept of the shuttle in low-earth orbit. This cutaway view shows the crew on duty in the flight deck. The huge payload doors are open as the orbiter has just released a payload into orbit. Also shown is part of the forward compartment below the flight deck. (Courtesy Rockwell International Space Division)

major structures such as the wings, mid-fuselage, vertical stabilizer and the titanium aft-thrust structure'' of Orbiter 103. Orbiter 103, named the *Discovery*, was originally scheduled for delivery in March 1982. Orbiters 105, the *Atlantis*, and 099, the *Challenger*, will follow.

Eventually, NASA plans call for a total of five shuttle orbiters to be built and in service. Perhaps five hundred flights will be made during the operational lifetime of the Space Transportation System, 1981–1992. (Courtesy Rockwell International Space Division)

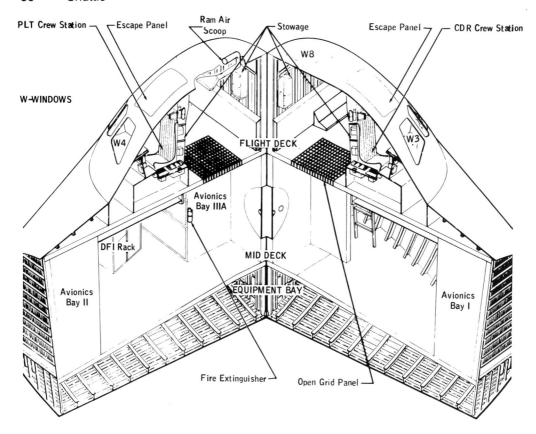

PLT Crew Station — ┌─Escape Panel   Ram Air   ┌─ Stowage   Escape Panel ┐   ┌─ CDR Crew Station
                                        Scoop

                                                    W8

W-WINDOWS

W4   FLIGHT DECK   W3

                              Avionics
                              Bay IIIA

        DFI Rack   MID DECK

Avionics   EQUIPMENT BAY   Avionics
Bay II   Bay I

                Fire Extinguisher ─┘   Open Grid Panel ─┘

Crew module. (Courtesy NASA)

Orbiter is as unlike Apollo as a DC-3 was unlike a Sopwith Camel, perhaps a slightly unfair comparison. (Apollo wasn't quite that primitive and the DC-3 wasn't quite that sophisticated.) Launch gravity, or g, forces for the shuttle are one-third that of Apollo: 3.0 g on launch and only 1.5 g on reentry. And orbiter is a shirtsleeve environment. There will be no bulky space suits or heavy flight gear worn inside an orbiter. The biggest difference of shuttle is the *method* of travel: launch like a rocket, maneuver in space like a spacecraft, and land like a conventional airplane.

The orbiter is divided into major sections: the forward fuselage (consisting of the upper and lower forward fuselage and the crew module), wings, mid-fuselage, cargo bay doors, aft fuselage, and the tall vertical tail. Most of it is aluminum with special heat protection for reentry.

There are six forward windows—the shuttle windshield—two overhead windows, and two rear-viewing cargo bay windows on the upper flight deck. There is also a window in the crew hatch which opens into the deck of the crew module. The six windshields are the largest pieces of space glass ever produced that are transparent enough to see through. They are each 45 inches diagonally and are made up of three panes of special glass. The outer and inner panes are .625 inch (1.5 centimeters) thick, while the center pane is 1.3 inches (3.3 centimeters) thick. The other windows in the orbiter are similarly constructed, including the small 10-inch (25.4-centimeter) window in the entry hatch.

*The specially constructed windows of the space shuttle. Astronaut Joe Engle can be seen in the cockpit of the **Enterprise** during a testing program in the Mojave Desert in September 1977. (Courtesy NASA)*

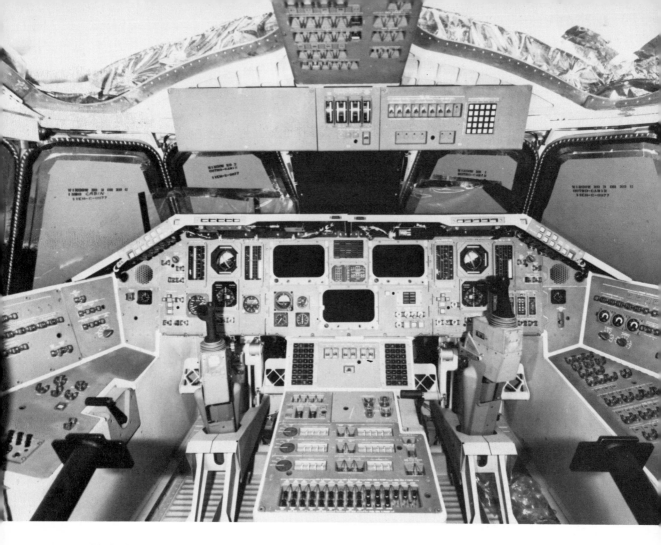

*Flight deck of Orbiter 101, the **Enterprise**. This is the primary forward-facing flight station, with dual pilot and copilot controls. An orbiter can be piloted, as can an aircraft, from either seat and by only one crew member in an emergency. (Courtesy NASA)*

Access to the crew module is through the orbiter hatch, a 40-inch (101.60-centimeter) side hatch in the mid-deck, which opens outward down 90 degrees when the orbiter is horizontal (as in the landed position) or 90 degrees sideways, if the orbiter is vertical (in launch configuration). The orbiter cabin is divided into three sections: the flight deck, the midsection, and the lower section.

From the flight deck—sometimes called the upper flight deck—the shuttle crew controls the launch, orbit maneuvering, and atmospheric reentry. Displays and controls in this section also control the integrated shuttle vehicle and the mission payloads. There is room for four crew members in the upper flight deck: the shuttle commander, the pilot, and two mission specialists from NASA.

For rescue, the shuttle crew will have two different systems. Pilots, copilots, and mission specialists will have the shuttle space suit, but passengers, payload specialists, will have the personal rescue system shown: a 34-inch diameter ball which contains its own short-term life support and a communications system. The ball is made of the same material as a shuttle space suit and has a small viewing port. (Courtesy NASA)

The shuttle space suit, called the extravehicular mobility unit. It is a two-piece suit with an upper torso of rigid aluminum and a life support system in a backback. The suits will be available in small, medium, and large sizes to fit passengers for the shuttle flights. (Courtesy NASA)

An artist's view of a shuttle extravehicular mobility unit. It has two pieces: an upper torso of rigid aluminum and a life support system in a backpack. The suits will be available in small, medium, and large to fit passengers for the shuttle flights. (Courtesy NASA)

The manned maneuvering unit will allow shuttle crew to range outside the cargo payload bay. The modular device will be stowed in the cargo bay and can be attached to the extravehicular mobility unit for extensive EVAs. It has electrical outlets for power tools, a portable light, cameras, and instrument monitoring devices. A crew member can fly with it to nearby structures and free-flying experiments. In the future, the manned maneuvering unit may help construct large space structures. (Courtesy NASA)

The midsection, or simply the deck, contains passenger seating, the living area, an airlock, and compartments holding electronics equipment. There is an aft hatch in the airlock, which gives access to the cargo bay of the space shuttle and the payload or payloads aboard.

For working outside the orbiter, or in the payload bay, the airlock is essential. It is equipped with handrails and handholds, and the orbiter is equipped with two heavy space suits, called extravehicular mobility units. Perhaps the most important feature of the airlock is that it can be located in several places for the flight: inside the mid-deck on the aft bulkhead, outside the cabin on the aft bulkhead, or on top of a tunnel-adapter fitting which connects the spacelab with the orbiter cabin. When a docking maneuver is planned, the docking module serves as the extravehicular activity airlock.

The airlock passages are designed with a straight-through passage. This makes it possible for equipment to be moved from inside the orbiter and the loose equipment storage in the mid-deck to the payload cargo bay. The airlock hatches are "D"-shaped. Even wearing bulky space suits, two crew members can carry a package 18 by 18 by 50 inches (45 by 45 by 127 centimeters) through the airlock.

The orbiter carries enough life support for two six-hour extravehicular activities, and life support is provided for one emergency trip.

The lower deck contains the environmental control equipment for the orbiter. The environmental control system supplies the flight crew and passengers with a comfortable and safe environment; maintains the proper atmospheric pressure, humidity, carbon dioxide level, temperatures; and removes odors from the cabin area. Access to the lower deck from the deck above is through removable floor panels.

The upper flight deck is organized into four basic areas: (1) two forward-facing, primary flight stations for vehicle operations, (2) two aft-facing stations, one for payload handling and the other for docking maneuvers, (3) a payload station for the management and checkout of active payloads, and (4) a mission station for orbiter internal power and communications, and connections between the orbiter and the payload.

The two forward-facing, primary flight stations have the usual pilot-copilot relationship with duplicated controls, so that the orbiter can be piloted from either seat, or operated by one crew man or woman in the event of an emergency. There are manual as well as automatic controls at each of the stations.

The payload handling station has the displays and controls for manipulating, deploying, releasing, and capturing space shuttle payloads. The crew member at this station can open and close the cargo bay doors, use the orbiter manipulator arms, operate lights and television cameras in the payload bay, and perform other duties. For monitoring the success of payload operations, two closed-circuit television monitors display video from the payload bay.

The rendezvous and docking station has displays and controls for rendezvous

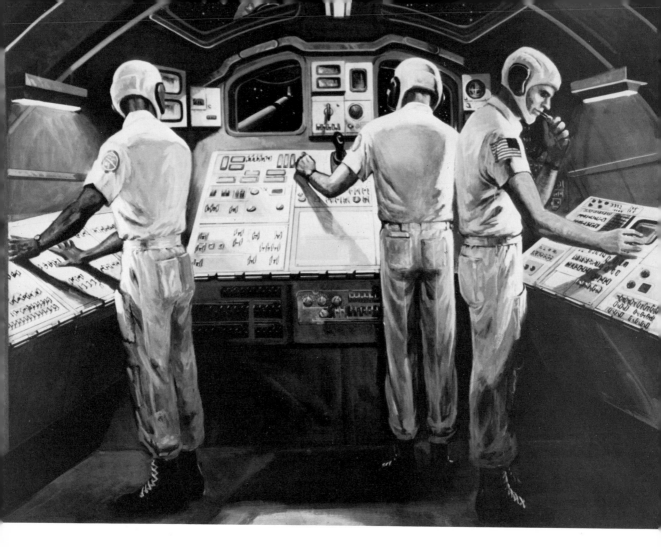

Payload handling is done by crew members at the aft, flight-deck payload station. (Courtesy NASA)

radar and rotation, hand controls for translation or lateral movement, and an attitude and direction indicator.

The payload station, just aft and to the left of the commander's chair, has a 21.5-square-foot (2-square-meter) surface area for installing controls and equipment for a specific shuttle payload—Space Telescope, spacelab, and planetary exploration spacecraft, among others. From this station can come payload power, monitoring, command, and control.

In the mission station, aft and to the right of the pilot's chair, are the displays and controls for managing the systems which are critical to the orbiter. There are caution and warning systems at this station to alert the crew to critical problems and potential malfunctions. From this station the crew member can monitor, command, and control attached and detached payloads.

Like an old whaling ship of the last century, the orbiter is full of nooks and crannies in which gear and equipment are stowed. About 150 cubic feet (4.2 cubic meters) in the crew compartment are used for loose equipment stowage. Most of this will be on the mid-deck. Each mission will be different, like old sailing voyages, so the amount and kind of extra gear aboard will vary from shuttle flight to shuttle flight.

The extra equipment stowage is limited by the size of the entry hatch and the structure of the mid-deck. Most of the loose payload aboard the orbiter will be stowed in standard containers attached to the deck walls and forward bulkhead.

Stowed in the crew compartment will be the crew's personal items. Like old sailormen, the crew will pack small bags with their own personal belongings for the trip. As a result of the great flap which occurred when Apollo astronauts carried with them unauthorized personal belongings (stamps and envelopes to cancel in space and increase their value), the shuttle crews will probably have to put up with security checks of their personal belongings.

Space Shuttle
Closed Circuit TV System

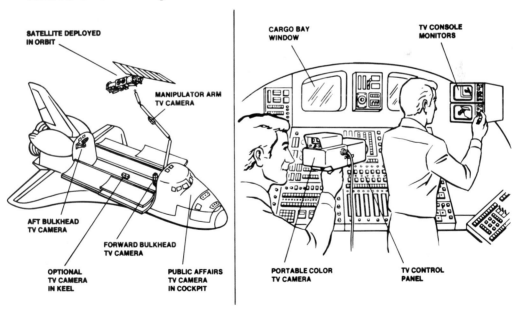

SATELLITE DEPLOYED
IN ORBIT

MANIPULATOR ARM
TV CAMERA

AFT BULKHEAD
TV CAMERA

FORWARD BULKHEAD
TV CAMERA

OPTIONAL
TV CAMERA
IN KEEL

PUBLIC AFFAIRS
TV CAMERA
IN COCKPIT

CARGO BAY
WINDOW

TV CONSOLE
MONITORS

PORTABLE COLOR
TV CAMERA

TV CONTROL
PANEL

RCA ILLUSTRATION

The orbiter crew keeps track of the payload operations and other mission functions with television cameras. (Courtesy RCA Astro-Electronics)

Most of the crew baggage will be things to make a trip seem shorter or less eventful, or possibly more eventful in the case of a long mission. Games which can be played in zero-g will probably be among the most popular items, along with books. Regular cards are impossible in zero-g because the cards would leave the table according to Newton's Laws of Motion. They would also be reluctant to stay in one place and shuffling could be extremely tricky: one slip and the old game of "fifty-two pick up" would have new meaning with fifty-two cards in various ill-defined orbits around the crew compartment.

Still, poker, hearts, bridge, or something similar is not entirely impossible in space. A deck or wall covered with Velcro material would make a table. Velcro has one face with little hooks and the other with loops. When put together, Velcro will hold something in place, but it can be removed easily without harming the surfaces. The cards would be covered with Velcro on each side, making it possible to put them on the cloth table. Shuffling would be tedious, but possible. As an alternative, card games might be possible with cards of extremely thin metal and a magnetic card table.

Chess and checkers are ideal for space voyages when magnetic boards and pieces are used. As a matter of fact, any game which has a board and moving pieces which can be magnetized is a possibility.

Books of all sorts will be a space shuttle regular, all probably in paperback to reduce weight. For extended missions, books would be a real relief from boredom: there are so many necessary scientific duties aboard an orbiter, it would be hard to get another crew member at the right time for a game of any kind.

Communication with the earth is better from orbiter, if that is possible, than it was from Apollo. If crew members wanted to, and NASA allowed it, there would be no problem patching in the communications aboard the orbiter with telephone lines on earth. A crew member could call anywhere, talk to anyone, even conduct business from near-earth orbit.

It is always possible that earth television will be piped aboard the orbiter with a full range of programs or a cable arrangement with old movies. There are as many television possibilities in space as there are in the average hotel room or jet airliner. An orbiter crew might even take videotape with them and regularly schedule some movies on a crew television in the shuttle. The irony of seeing a rerun of *Star Trek* and its *Enterprise* aboard a *real Enterprise* high above the earth is slightly frightening, but nevertheless entirely possible.

Stereo, either in the form of piped stations from Houston or cassettes carried aboard the orbiter and played in space, will no doubt go into shuttle at some point. The bunks aboard the orbiter would be equipped with an earphone system much the same as used on aircraft.

One of the main recreations aboard an orbiter probably will be sightseeing. The earth will be an everchanging pattern of blues and browns, with gray and white sweeps of cloud. At night familiar cities can be picked out by their lights on the dark surface of the planet. The stars will be infinite and the familiar boundaries of constellations as seen from earth will be difficult to locate. The moon, the planets, the great sun will all be visible to orbiter crew members in their off-hours.

And if, as was the case with *Skylab,* some pets go along, there is always the fun of watching them try to adjust to weightlessness and the other conditions of space. Arabella, the first space spider, went along in an experiment aboard *Skylab* on the second manned mission and was an endless source of amusement. She and her sister Anita wobbled around at first and then settled down to spinning webs as usual. The webs from the two ordinary, Cross spiders did not differ greatly from those spun on earth. *Skylab* also carried a complement of minnows—Mummichog variety—which showed interesting reactions to zero-g. The minnows carried into space as young adults swam in tight loops, while those which hatched from eggs laid on earth swam in normal minnow fashion.

But mostly there will be work and more work. It won't be possible to just wander back to spacelab (if it is a spacelab mission) and see what's going on—spacelab's life support system is designed for operators *only* except during the shift changes. And, as the space shuttle becomes more routine, fewer of the off-duty hours will be consumed by obligations to media television and radio and curious inquiries from the ground.

Exercise aboard the orbiter will be necessary for most missions over one day long. The adjustment to zero-g is somewhat comparable to moving from a very low altitude environment to a high altitude one or vice versa, except the physiological changes are more pronounced. During the initial period of weightlessness, there is a profound loss of stamina and physical reserves. There is a jet lag as the body tries to adjust to the new conditions.

Such exercise hardware as abbreviated bicycles, rope and handle exercisers, and a treadmill device were used aboard the extended *Skylab* missions. For long duration shuttle missions, similar devices would probably be used, but as there is less room for them, they would have to be more compact.

Then there's the inevitable housekeeping: a small hand-held vacuum cleaner can be used to remove unwanted particles from the air and to clean the galley and crew's quarters. Trash will be disposed in the compact galley.

There are three bunks aboard the orbiter, and for those who don't want a bunk, a sleeping bag is available. The bunks have draw-curtains and are fitted with restraining straps; otherwise a completely relaxed sleeper would drift out of the bunk and into the crew module, floating on whatever air currents happen to be moving.

The sleeping bags are multilayered and can be hung on a frame against the crew compartment wall. When a shuttle astronaut is ready for sleep, he or she zips

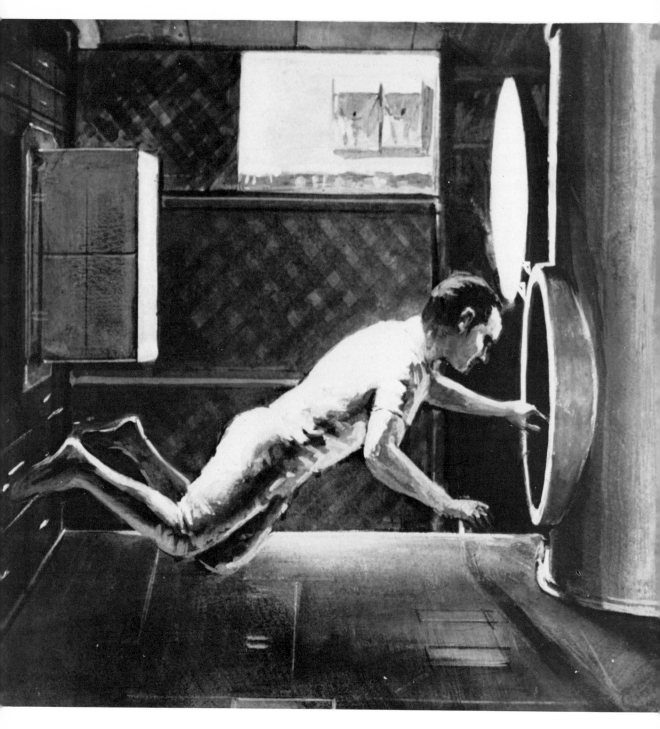

The crew and passengers of the orbiter occupy a two-level cabin at the forward end of the orbiter. This crew member is dozing on the lower deck. (Courtesy NASA)

open the sleeping bag, puts in some blankets, slides in, and then zips up. With a body strap around the sleeping bag, and the zipper closed, the bag is comfortable and there is no fear of floating around.

There is no problem with sleeping in zero-g space: there is no up or down so astronauts and passengers can sleep in any position. Without gravity it makes no difference whether your head is pointed toward the floor, a wall, or the ceiling.

Personal hygiene aboard the space shuttle is not quite up to the standards set by *Skylab,* which had much larger crew quarters. No shower is available, nor are there towels and wash cloths. The personal hygiene facility aboard shuttle, as it is euphemistically called by NASA, consists of plenty of wet wipes—moist towelettes about the same size and type as those given out by chicken eateries for their customers. It is possible to take a sponge bath with these, but it isn't easy; and the towelettes tend to float around, creating the possibility of an astronaut entering the crew compartment chasing after parts of his or her bath.

The shuttle toilet facilities are similar to those aboard *Skylab.* The main consideration is a strap for the contoured seat (to avoid departure a la Newton's Law of Motion) and foot holds. In the absence of gravity, vacuum flow separates wastes from the body.

Changing into clean uniforms can be accomplished easily in zero-g since the clothes slide on and off in peculiar ways with no gravity. The bunks, with their draw curtains, can provide privacy for uniform changes. Space shuttle living conditions probably most closely resemble a cross-country train ride, in a Pullman car—not as elegant as a hotel, but not as primitive as a 3000-mile journey by coach seat.

For those who want to change their appearance, even for awhile, a trip on the orbiter is a good idea. Generally, passengers on a lengthy mission could count on adding an inch in height (due to stretching of the vertebrae in the absence of gravity) and losing considerable weight, or at least becoming much slimmer from the redistribution of body fluids from the lower to the upper extremities.

Living and dining aboard the space shuttle will not be palatial, but they will certainly not be completely spartan. Menus for the shuttle have been set at 3000 calories per day. During Apollo, meals were set to give 2800 calories a day. And there are further evidences that space shuttle is the latest in space comfort. Shuttle crew members who like to put catsup on everything can do so. Those who can't make it to sleep without a cookie can have one, along with a choice of one of their favorite drinks.

In the early days of the space program the problem of eating and food never arose—the flights were too short. By the time Gemini flights were getting to the ''hungry'' stage, the astronauts were supplied with what might be called ''entree a la toothpaste tube.'' Those were the days when crumbs floating around a zero-gravity cabin had to be recaptured before they got into the instruments and ruined something.

Following, and often in addition to, the toothpaste-tube food for early space flight was a selection of lukewarm, dehydrated cubes or freeze-dried items. These very early space meals were consumed, often reluctantly, amid considerable groaning and complaints to Houston.

The long Apollo flights brought a considerable improvement in dining. There were new forms of food preparation, heating, and cooling, and new standards of palatability. By *Apollo 8,* the astronauts were surprised with a home-cooked Christmas dinner—pieces of turkey with most of the usual trimmings. And best of all, the meal was eaten with a spoon instead of being

This compact storage container holds three meals a day for three days for shuttle crew members. (Courtesy NASA)

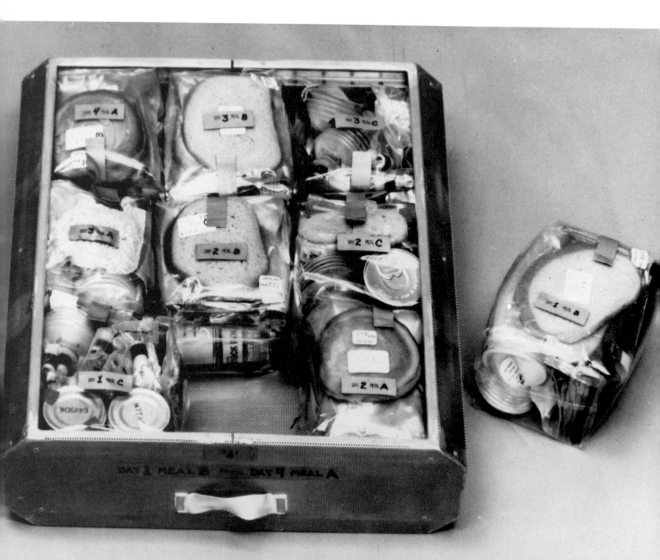

GALLEY DETAILS

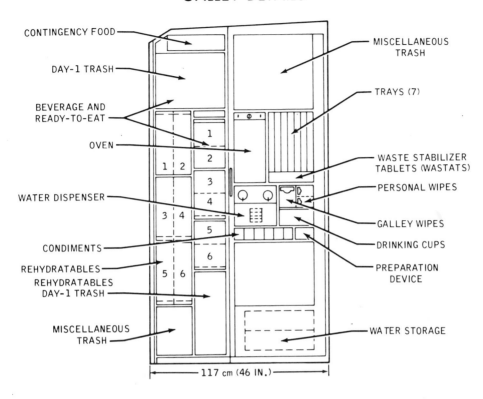

CONTINGENCY FOOD

DAY-1 TRASH

BEVERAGE AND
READY-TO-EAT

OVEN

WATER DISPENSER

CONDIMENTS

REHYDRATABLES

REHYDRATABLES
DAY-1 TRASH

MISCELLANEOUS
TRASH

MISCELLANEOUS
TRASH

TRAYS (7)

WASTE STABILIZER
TABLETS (WASTATS)

PERSONAL WIPES

GALLEY WIPES

DRINKING CUPS

PREPARATION
DEVICE

WATER STORAGE

|←——— 117 cm (46 IN.) ———→|

(Courtesy Rockwell International Space Division)

squeezed from a container. In the case of *Apollo 8,* the turkey and gravy was a thermostabilized product packaged in a laminated foil pouch. This event led to the development of the spoon-bowl plastic pack for rehydratable foods. Water was injected through a valve, and the contents were spooned out through a zippered opening.

When *Skylab* came around, the astronauts' grumbling about the food became a low but distinct murmur, instead of a howl. Each *Skylab* dweller received 4.4 pounds (1.9 kilograms) per day and, for the first time, there were freezers, refrigerators, and warming trays in space. Except for the drinks, which were taken from a collapsible dispenser, the meals were packaged in aluminum pop-top cans, which were put into the food warmer-retainer tray from which they were eaten.

By the time of *Apollo–Soyuz,* the Russians were eating ethnic foods packed in tins and aluminum tubes. There was a small heater aboard the spacecraft, and meals were often to the cosmonaut's order: meat, or meat paste, bread, cheese, soup, dried fruit and nuts, coffee, and cakes.

But the space shuttle—that is elegant space dining. There is a real oven in the galley and, hopefully, a refrigerator later, when space and power supply permit. The meals, unlike *Skylab* cooking, will be *hot*. In *Skylab,* with its less-than-normal air pressure, the foods never quite became hot, for the same reason water will not get extremely hot at a high enough altitude on earth.

The shuttle oven, on the other hand, operating in a cabin maintained at

Gravity-resistant food tray. Portions are held by Velcro tape and plastic envelopes are held down by springs. (Courtesy NASA)

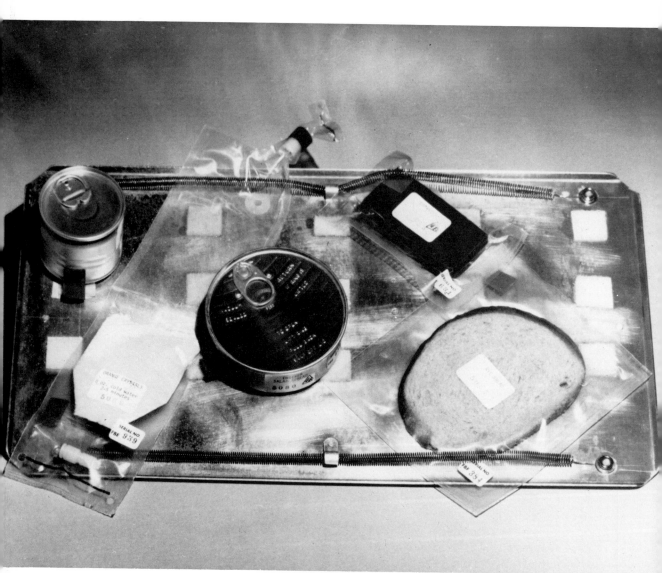

standard, sea-level atmospheric pressure, can heat meals to 185 degrees (85 degrees Celsius) and hold the heat level to 150 degrees (65 degrees Celsius) for warming or reheating. And it can heat foods in a vast variety of containers made from different materials, and in odd sizes and shapes.

The "roomy" shuttle galley has a pantry, oven, dishwasher, hot and cold running water, and even a dining table. A meal can be assembled by a crew member in four or five minutes, and heated and ready to eat in about an hour. Wet wipes are used for cleaning up the trays and the utensils. Crew members will be responsible for their own cooking and cleaning up—unless someone wants to be cook for a day.

Aboard shuttle there will be seventy-four kinds of foods and twenty kinds of drinks available to the crew. There will be a different menu for each of six days. On the seventh day the food cycle repeats. For those who don't want to eat at the galley dining table, the food trays can be held in the lap or stuck on a wall with special fastenings, even in zero gravity.

A typical daily menu aboard space shuttle might start with a breakfast of orange drink, peaches, scrambled eggs, sausage, sweet roll, and cocoa. Lunch might consist of cream of mushroom soup, ham and cheese sandwich, stewed tomatoes, banana, cookies, and tea. The big one, dinner, reads like something from an earthside restaurant: shrimp cocktail with sauce, beefsteak, broccoli au gratin, strawberries, pudding, cookie, and cocoa. Some of the shuttle foods aren't even special items. The breakfast bars are right off a supermarket shelf.

Best of all, the scrambled eggs do not taste like C-rations, there may be ice cream—not as yet planned on the menu but a favorite of *Skylab* astronauts—and there will be no cold potatoes. Cold potatoes were voted the most repulsive by all the astronauts of all space missions.

Electrical power for the space shuttle and the various payloads is generated by three fuel cells that use cryogenically stored hydrogen and oxygen. These fuel cells are situated in the forward end of the orbiter mid-fuselage. The amount of hydrogen and oxygen for the cells normally carried aboard orbiter is enough to produce 1530 kilowatt hours of energy, including the 50 kilowatt hours needed by a typical payload for a typical seven-day mission into space. When missions are extended to thirty days or there are special requirements for added power, the orbiter will carry extra fuel kits. Each of the extra kits is good for 840 kilowatt hours of energy.

The chemical conversion of hydrogen and oxygen to obtain electrical energy produces a by-product which is—wonderfully—water. Thus, all, and more, of the water needed for the orbiter crew comes from the process which produces the power for the craft.

The range of potential shuttle missions requires the orbiter to accommodate very different payloads. To take care of the variety, the orbiter payload bay has attachment points along the sides and floor of the 60-foot (180-meter) cavern. The cargo bay also has attachment points and facilities for extra equipment for the crew, additional power kits, and small payloads. Included in payload cargo-bay equipment is a special docking module designed after international standards, so that international use of shuttle in the next twenty years will be a relatively simple operation.

The orbiter's twenty communications antennas are mounted flush on the forward fuselage. The entire system allows the orbiter to provide direct voice, command, telemetry, and television communications with the ground crew, crew members on extravehicular activity, and detached payloads. Orbiter can link with any network of ground stations and communications satellites.

"Houston, *ici* Orbiter 102. *Nous avons ouvert les portes la soute, et nous sommes prêts pour les operations de largage.*" An imaginary conversation with Houston from a French mission specialist, who is about to use one of the most interesting devices aboard shuttle. The payload specialist has just informed Houston that the payload doors are open and he is "go" for deployment. Deployment (*"les operations de largage"*) uses the remote manipulator arm, which functions like a human arm and is operated from the upper flight deck on a control panel. To move the tip of the arm through space, the operator pushes a control knob with his left hand, back and forward, up or down, to one side or another, according to which way he wants the arm to go.

To point the end of the arm in pitch, yaw, or roll, the operator rotates a control stick in his right hand, a stick similar to a helicopter joy stick. One of the five computers aboard orbiter signals the manipulator arm and makes sure the motion of the motors in the arm corresponds with the instructions from the operator's control panel. Visual feedback to the operator comes in through the overhead and aft-facing windows looking out over the huge cargo bay.

The manipulator arm is about as long as two telephone poles laid end to end. In space it can handle a load as large as a bus. Since it is supposed to function like a giant human arm, it is not unlike one. It is attached to the orbiter by a shoulder joint, it is hinged at its mid-point for an elbow, and at the end of the arm is a hand, moving in a wrist joint.

The arm will be used to remove new satellites which have been flown up to space in the space shuttle cargo bay, to reach out and drag in old satellites which are to be repaired, and to shift cargo of any kind within the cargo bay.

With the cargo bay, the manipulator, and the hundreds of displays, controls and read-outs, the orbiter and its crew can do operations in space which were only dreamed of in the days of expendable launch vehicles. It is a complex spacecraft—half rocket, half airplane. And with it, the next twenty years in space will be both routine, and infinitely exciting with possibilities.

5

LAUNCHES, LANDINGS, AND IN BETWEEN

The space shuttle's payload bay contains a communications earth satellite among the several scientific wonders which crowd its cavernous cargo space. Far above Cape Canaveral where they were launched, the crew of the space shuttle begins maneuvers to release the satellite at the place and time which the computers in Houston and other distant cities have decreed. Fire a vernier engine, there just so; fire attitude control thrusters now for this long; maintain attitude. The computers chatter anonymously to each other in non-time and non-words.

Silently the small flames lick outward from nose and tail, and the universe as seen from the space shuttle begins to wheel about. The final computer whisperings tell the story of completed maneuvers. Complex words and concepts reduced to minute electrical charges have turned a great machine high above the earth around 180 degrees to point its belly toward the earth and its cargo bay to the dark and black of space. What was it Arthur C. Clarke said—"Any sufficiently advanced technology is indistinguishable from magic." What an astonishing display of demons would some seventeenth century man think this was, this silent firing of gasps of dragon flame, this maneuvering procedure in a metal projectile alone in a vacuum unfit for man.

America's first rocket port, as early science fiction stories called the launch sites of spaceships, was White Sands Proving Ground near Alamagordo, New Mexico. It was located on a flat, immense, unbroken plain in a part of the southwestern United States which seemed to be totally barren, empty, and desolate. But from the viewpoint of firing rockets, it was anything but desirable, despite the apparent emptiness.

The cities of El Paso, Albuquerque, and Santa Fe were uncomfortably in range of a runaway rocket, especially since rockets were expected to become increasingly more powerful. Even worse was the closeness of Mexico and the potential for an international incident.

When one of the captured V-2s from World War II, which were launched at White Sands in the late 1940s, departed for the Mexican border on one flight, it was quickly decided that a new location would have to be picked for a major rocket-testing facility. The errant V-2, fortunately, did no damage to the Mexican countryside, livestock, or humans, and impacted near the border town of Juarez.

The search for a new launch site ended on a sandy promontory on the Atlantic side of Florida in 1950. Originally established as the Air Force Proving Ground at Cocoa, Florida, it had the added advantage of Patrick Air Force Base only fifteen miles (24 kilometers) to the south. Cocoa, more properly known as Cocoa Beach, is a lovely little community by the shores of the Atlantic Ocean sitting at one end of a main street, the other end of which passes through the once tiny hamlet of Cape Canaveral.

Modified V-2 rockets were fired from the site at Cape Canaveral as early as 1950. There is a story, probably apocryphal, that the earliest launches at the Cape were from a converted tennis court—Cape Canaveral was once a resort—using a Sherman tank hulk as a block house to protect the scientists and rocket engineers.

Dotting the Cape are the permanent launch pads which have figured in the history of space travel and exploration. Complexes 5, 6, and 26 now are the Air Force Space Museum. From these launch pads, the early Jupiter and Redstone rockets were flown, including the Mercury flights. From Complex 17 came the launches of scores of earth satellites including Telstar, Explorer, Tiros weather satellites, and others. Further along the sandy beaches at the easternmost point of Cape Canaveral are the pads for the Atlas-Centaur rockets which launched Surveyor to the moon, followed by the Atlas-Agena launch complex which sent Ranger moonward, and Complex 19, which was launch site for the Gemini missions.

The Cape is largely separated from the mainland by two bodies of water and an island. Inland from Cape Canaveral proper, across the Banana River, is Merritt Island, which in turn is separated from the mainland by the Indian River. Along the mainland road which parallels the banks of the Indian River are the communities of Daytona Beach, Titusville, and Melbourne.

Actually, the Cape has two divisions: NASA, in the form of Kennedy Space

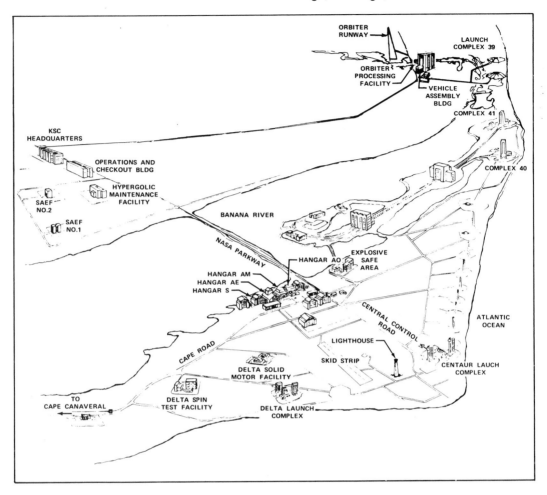

KSC FACILITIES AND SERVICES

(Courtesy Kennedy Space Center)

Center, controls civilian space programs; the Cape Canaveral Air Force Station handles military launches. The Air Force part of the Cape is reached by coming north from Cocoa Beach and the town of Cape Canaveral, along NASA Causeway East across the Banana River from Merritt Island, and by the Cape Road from the north, which sits atop a tiny spit of sand connecting the actual Cape with a thin, sandy extension of Merritt Island.

NASA built up its part of the Cape as the manned space program increased in size. The technical dividing line between Cape Canaveral Air Force Station and

The other shuttle landing site, Vandenberg Air Force Base, California, is shown in this artist's conception. (Courtesy NASA)

Kennedy Space Center is just south of Launch Complex 41, which sent Viking on the way to Mars and Voyager out to Jupiter and Saturn. Technically, on the Merritt Island side in the middle of marshes are the Apollo/Saturn launch areas, Complex 39, pads A and B. It was from here that the flights to the moon were launched, and it is these same pads, much modified, that the space shuttle will use.

In 1963, the entire area was renamed Cape Kennedy, but that slowly fell into disuse, and in 1973 it reverted to its original name of Cape Canaveral, home of the Kennedy Space Center and the Air Force Test Station. But nearly everyone in the space business calls it simply "the Cape."

It is an ideal spaceport, bordered by the immenseness of the Atlantic Ocean where any mistakes, and there have been very few, make no difference. First and second stages of rockets drop into the ocean downrange where no one could possibly be harmed.

The other American launch site for rockets is Vandenberg Air Force Base in California. Space shuttle will also be launched and land at Vandenberg. The Western Test Range, as it is called (the Cape is sometimes referred to as the Eastern Test Range by purists), also provides immense ocean distances over which stages can splash down and rockets can blow up without harm to inhabitants.

The orbits which can be reached by the shuttle are determined by the physical geography of the North Americas. The initial portions of a shuttle flight will not pass over land for reasons of safety. Therefore, from the Cape the northernmost launch azimuth is fixed by the southern portion of Labrador and on the south by the Bahamas. For Vandenberg, the limits are the Hawaiian Islands and portions of Mexico. Vandenberg can be used for launches into polar orbit and other similar inclinations, and Kennedy Space Center will be used for launches into more or less equatorial orbits.

The launch site also affects the payload-carrying capacity of the shuttle. A due-east launch from the Cape gets a launch assist from the earth's easterly rotation of 1000 miles per hour (1600 kilometers per hour). The payload can be nearly 65,000 pounds (29,484 kilograms), the most quoted figure for the shuttle's payload. Very roughly, 65,000 pounds could be fifteen Lincoln Continentals. On Vandenberg launch at its best, however, the earth's rotation is a neutral factor, meaning a payload capability of 40,000 pounds (18,144 kilograms). The worst Vandenberg launch—the most westerly one, where the launch is in a direction opposite to the earth's rotation—allows only a payload of 32,000 pounds (14,515 kilograms).

Vandenberg is not expected to be available until the shuttle becomes operational in the 1980s. A few of the first orbital test flights will land at Edwards Air Force Base, California, instead of on the shuttle runway at the Cape, but Edwards is not to be confused with Vandenberg.

Operational Flight Ground Turnaround Time

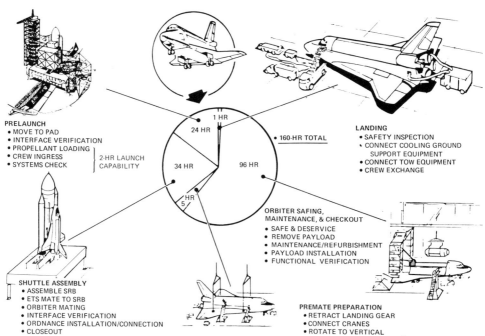

PRELAUNCH
- MOVE TO PAD
- INTERFACE VERIFICATION
- PROPELLANT LOADING
- CREW INGRESS
- SYSTEMS CHECK

2-HR LAUNCH CAPABILITY

LANDING
- SAFETY INSPECTION
- CONNECT COOLING GROUND SUPPORT EQUIPMENT
- CONNECT TOW EQUIPMENT
- CREW EXCHANGE

1 HR
24 HR
• 160-HR TOTAL
34 HR
96 HR
HR 5

ORBITER SAFING, MAINTENANCE, & CHECKOUT
- SAFE & DESERVICE
- REMOVE PAYLOAD
- MAINTENANCE/REFURBISHMENT
- PAYLOAD INSTALLATION
- FUNCTIONAL VERIFICATION

SHUTTLE ASSEMBLY
- ASSEMBLE SRB
- ETS MATE TO SRB
- ORBITER MATING
- INTERFACE VERIFICATION
- ORDNANCE INSTALLATION/CONNECTION
- CLOSEOUT

PREMATE PREPARATION
- RETRACT LANDING GEAR
- CONNECT CRANES
- ROTATE TO VERTICAL

(Courtesy NASA)

In almost all the civilian flights of shuttle, the Cape will be the launch and landing point. It has been extensively changed to meet the new assignment, and it is the key to the world of the space shuttle. The facilities at Kennedy Space Center allow a shuttle flight to be completely recycled in about 160 hours, or two weeks turnaround time.

When an orbiter lands from a mission, the ground turnaround cycle begins. On the runway, the preliminary servicing is accomplished, together with appropriate safety procedures. Then the orbiter is towed away to an area where any Department of Defense payloads are removed, and pressurized containers and fuel cells are vented. Hazardous materials are removed from the shuttle.

The orbiter is then moved to a maintenance area where it is inspected. The insulation is renewed on the orbiter skin which was lost after the reentry into the earth's atmosphere following the previous flight.

The new payload is installed in the orbiter in a maintenance facility, although certain payloads—hazardous, time-critical, and Department of Defense—will be put aboard the orbiter on the launch pad, using the payload changeout room at the

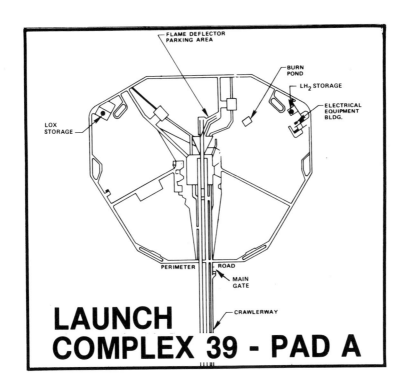

LAUNCH COMPLEX 39 - PAD A

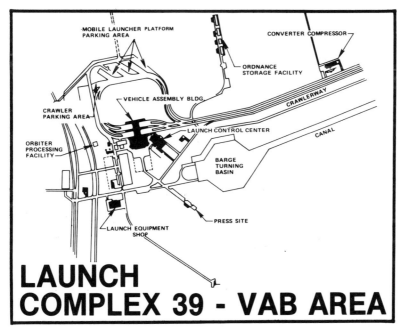

LAUNCH COMPLEX 39 - VAB AREA

(Courtesy NASA)

The Orbiter Processing Facility at KSC. In this building, the shuttle will be serviced after landing from a mission. It is northwest of the Vehicle Assembly Building at Kennedy Space Center. (Courtesy NASA)

Looking south, the orbiter's landing strip at Kennedy Space Center, Florida. This view was taken from an altitude of 1500 feet (457.2 meters). (Courtesy NASA)

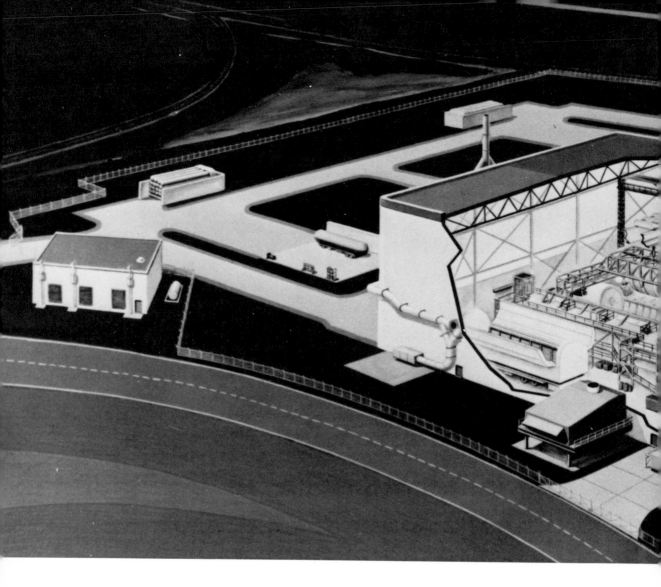

pad. All of the shuttle equipment, which has been serviced in another part of Kennedy Space Center, is installed in the shuttle. The orbiter is then moved to the Vehicle Assembly building.

When the orbiter has been assembled and checked out, it is mated to the already checked out external tank and the refurbished solid rocket boosters. Everything which has been put together is checked again.

The entire space shuttle is then rolled out, connected to the facilities at the launch pad, and checked for launch readiness. Countdown operations, loading of

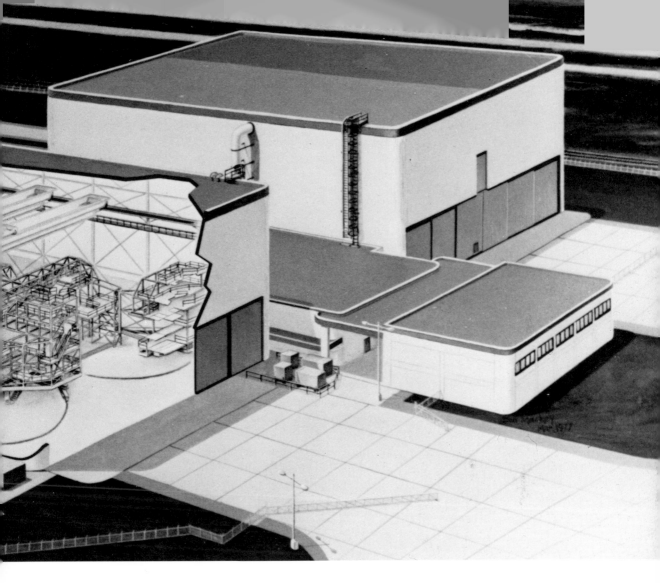

An artist's view of the Orbiter Processing Facility at Kennedy Space Center. (Courtesy NASA)

fuel, and the entrance of the next flight crew completes a shuttle cycle, which ends with the final automatic countdown sequencing, and a new lift-off for near-earth orbit.

One of the Kennedy Space Center's more impressive space shuttle facilities is the Orbiter Landing Facility. Most people would call it a runway, which it is, and one of the longest concrete runways in the world. The shuttle will have plenty of room on this 15,000-foot (4.5-kilometer) strip which is 300-feet (91-meters) wide. This is about the length of the longest runways at Kennedy International, and considerably longer (three quarters of a mile) than the longest at Dulles or

O'Hare. At each end of the runway there is a 1000-foot (300-meter) overrun section. The runway is located northwest of the giant Vehicle Assembly Building, perhaps the most noticeable landmark at Kennedy Space Center.

At the end of a two-mile (3.2-kilometer) towway from the landing strip is the Orbiter Processing Facility, otherwise known as a hanger. In this clean room environment, the ordnance and any leftover fuels are made safe. Also at the hanger, the flight and landing systems will be refurbished, old payloads removed, and new payloads installed.

The landing strip and the processing facility are the only completely new landmarks at the Cape. The remaining space shuttle preparations at the Cape have involved modifications of existing buildings and hardware originally designed for the Apollo and *Skylab* missions.

Out on launch pad A of Complex 39 the major change is a fixed, shuttle-service access tower, a water sound-suppression system to protect the shuttle crews and delicate payloads from acoustical damage during liftoff, and the payload changeout room. The payload changeout room allows loading and unloading of payloads at the launch pad.

The PCR is the white room structure mounted on a semicircular track extending from the service and access tower. Prior to a launch, it is retracted along its track to a parking site on the pad. Pad B of complex 39 will have similar modifications so it, too, can be used for launching the space shuttle.

The monster crawler-transporters, the huge-tracked vehicles which were used to move the Apollo-Saturn V rockets around the launch complex, will also be used for the space shuttle. The crawlers have been modified to carry the completely assembled space shuttle, along with its mobile launch platform, between the Vehicle Assembly Building and the two launch pads of Complex 39.

The firing rooms at the Cape, familiar to most of those who watched television coverage of the Apollo missions, have been modified for the space shuttle by adding a highly automated launch processing system developed for shuttle checkout and launch. Otherwise, Launch Control will look about the same as it did during Apollo. One main difference will be the number of people in the launch control: the lunar flights needed 450 technicians, engineers, and managers to get off the ground, and a twenty-eight hour countdown. Shuttle will need two-and-one-half hours and forty-five people.

The great Vehicle Assembly Building, so tall that it is said that there are occasional weather patterns, even storms, inside has also been modifed for the years when the space shuttle is operational. It will be used to stack and integrate flight components and to store the giant external tank of the space shuttle and the solid rocket boosters. In another part of the building, parts of the solid rocket boosters will be replaced with new equipment.

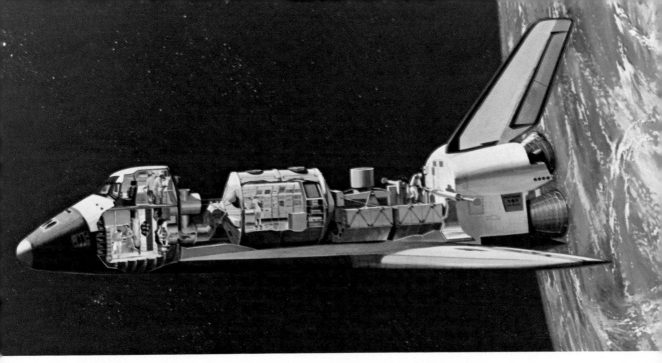

A cut-away view showing orbiter flight with Spacelab. (Courtesy NASA)

Shuttle rescue mission EVA. The astronaut in the foreground, equipped with a manned maneuvering unit, has carried a crew member in a rescue ball from an orbiter craft to shuttle. (Courtesy NASA)

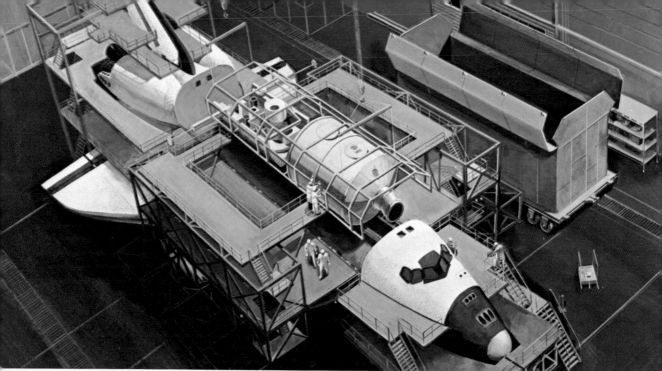

Payload installation operations in the Vehicle Assembly Building at Kennedy Space Center. (Courtesy Rockwell International Space Division)

Inside the Vehicle Assembly Building, the orbiter is mated with the external tank and the solid rocket boosters. (Courtesy NASA)

Engineers inspect aluminum slosh baffles that line the interior of the external tank. Installed inside the liquid oxygen portion of the tank, they diffuse the movement of the heavy fuel during launch. (Courtesy Martin Marietta)

External tank being delivered by barge to NASA Material Space Technology Laboratories in Mississippi for testing. (Courtesy Martin Marietta)

Astronauts are trained in the flight deck of a full-scale shuttle orbiter mock-up. (Courtesy NASA)

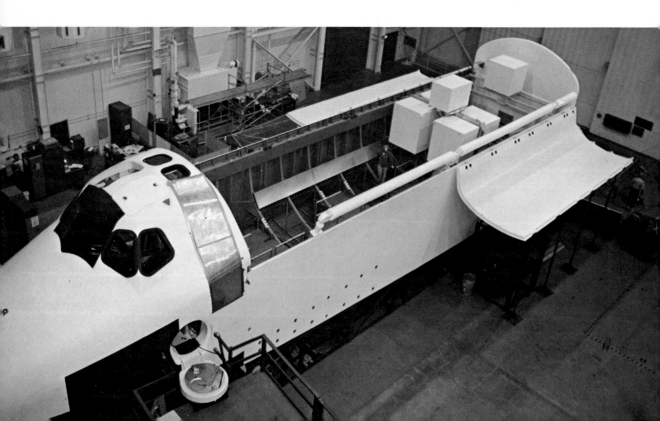

The space shuttle Service and Access Tower at Pad A (center left). It is constructed of elements of the Mobile Launcher 3 umbilical tower. (Courtesy Kennedy Space Center)

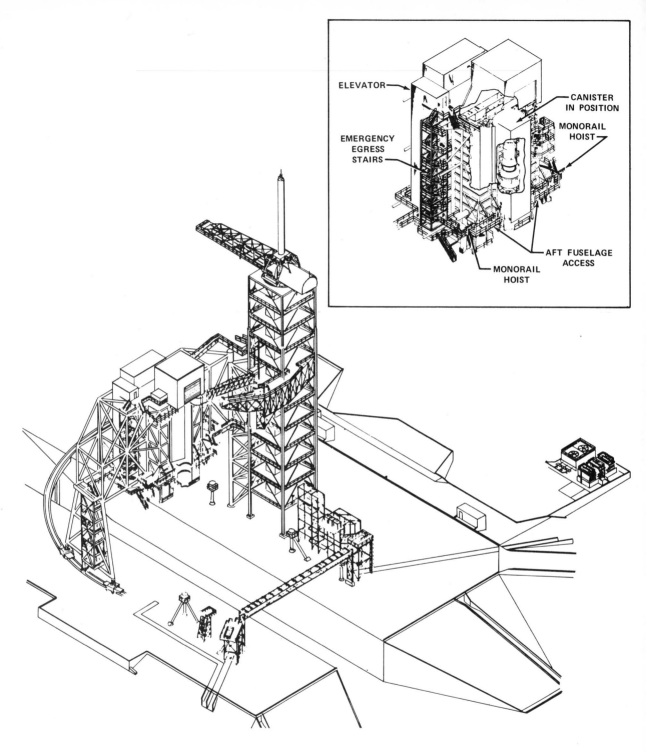

ELEVATOR

CANISTER
IN POSITION

MONORAIL
HOIST

EMERGENCY
EGRESS
STAIRS

AFT FUSELAGE
ACCESS

MONORAIL
HOIST

Payload changeout room in the position for payload transfer into the orbiter. No space shuttle is on the launch pad so all structures can be seen. (Courtesy NASA)

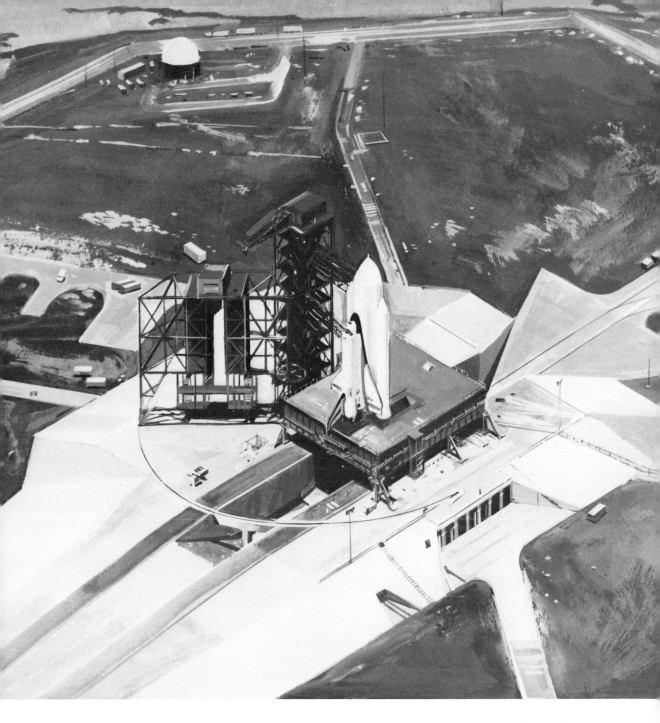

Space shuttle on the launch pad. Beneath the pad is the enormous flame trench which deflects rocket exhaust from the pad. (Courtesy NASA)

*The shuttle goes toward the launch pad aboard the giant crawler-transporter, which was also used during the **Apollo** lunar landing program. (Courtesy NASA)*

Spacelab will have its facilities in the old operations and checkout building which once accommodated the Apollo spacecraft before they were mated to the Saturn V launch vehicle.

Also at Kennedy Space Center will be buildings for checking out the orbiter's forward reaction control system and the aft propulsion systems after each mission. There is also a parachute handling area which will be used for work with the large parachutes which are used on the solid rocket boosters for their landing in the ocean. The chutes will be washed, dried, stored, and packaged for reuse.

Vertical space shuttle payloads, such as the interim upper stages and the solid spinning upper stages, will be processed and checked out in a building called the vertical processing facility, prior to installation in the shuttle orbiter cargo bay at the launch pad. The building, the former Apollo spacecraft assembly and encapsulation facility, has been highly modified for its new role.

South of Kennedy Space Center on the grounds of the Cape Canaveral Air Force Station, there will be one building which will be used for the space shuttle. A hanger will be modified to be used as a recovery and disassembly building for servicing the solid rocket boosters.

At the rear of the recovery and disassembly facility on the Air Force Station will be a barge slip. Solid rocket boosters which have been recovered from the ocean by ships will be sent by barge back to the Cape for rebuilding.

The operation at the Cape is complex. The giant external tanks come by barge from New Orleans where they are built, around the tip of Florida to Kennedy Space Center. The solid rocket boosters will be shipped to the Cape until a stockpile of recovered boosters makes that unnecessary or infrequent. The orbiters of the space shuttle will arrive from California on the back of a specially modified 747 aircraft. Parts which need reworking between flights will be shipped by rail to the Cape, as will propulsion segments for the boosters.

But apart from the fascination of hardware, the true ingenuity of the space shuttle and the complex devices which make it possible is the world of payloads: the Space Telescope, spacelab, the multitude of satellite possibilities, and the endless potential for space manufacturing and the civilization of space.

The first orbital test flights will feature payloads such as the Long Duration Exposure Facility and the Synchronous Communications Satellite. The operational flights, scheduled to begin in the summer of 1980, will feature in the first missions weather satellites, communications satellites for India, Indonesia, and other countries, and the Tracking and Data Relay Satellite. On tentative schedule are the launch of Spacelab-1 in December 1980 and Telsat Canada for early the next year. In late 1981, Spacelabs 2 and 3 will be launched aboard the space

The solid rocket boosters are recovered from the ocean and towed back to Kennedy Space Center where they are refurbished and used on other shuttle flights. (Courtesy NASA)

shuttle. The first interplanetary exploration craft to be launched with the shuttle will go into space in 1983–4.

The Tracking and Data Relay Satellite System plays a major role in the planning for the space shuttle and it is for this reason that it will be launched early in the shuttle's first flights. It consists of four specialized communications satellites in geosynchronous orbit and a ground terminal at White Sands, New Mexico. One of the satellites will be used by Western Union as a domestic communications satellite and one will be a spare. The remaining two will provide NASA tracking, command, and data relay.

The operational phases of the space shuttle will be supported by the satellite system—it will relay data and commands directly between Mission Control at Johnson Space Center and the space shuttle. It can also provide nearly continuous coverage for all near-earth orbiting satellites.

Because the Tracking and Data Relay System can transmit data very quickly and uses a wide band of frequencies, it can easily handle the Spacelab and similar missions. With its global coverage, this satellite system will increase nearly six-fold the time available for high-rate transmissions on NASA systems.

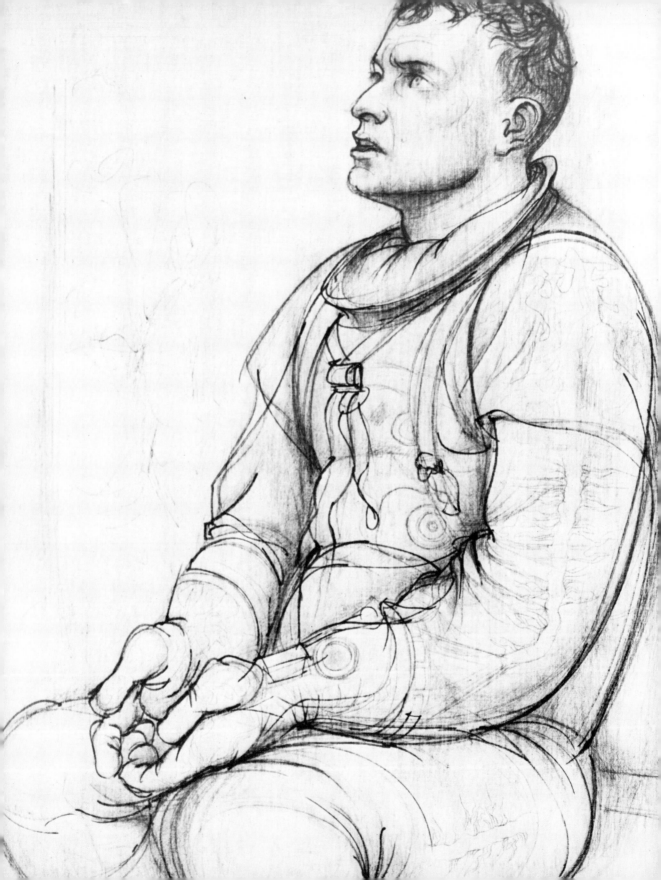

Look at him! There goes that crazy man who wants to fly.

a waiter to rocket pioneer Hermann Oberth
while pointing out Count Zeppelin in a restaurant

6

THE NEW ASTRONAUTS

In the orbiter, a man checks the continuing computer read-outs of position. A woman in slim uniform and uncoiled weightless hair makes adjustments to controls which have giant machine analogs in the cargo bay of the space shuttle. Twisting her thin arms and slender hands with sure, smooth motions, she causes the powerful mechanical shoulder joint, elbow, wrist, claw out in the cargo bay to fit into a deployment section of an earth satellite.

The satellite is left without power from the orbiter which has been keeping it alive since launch. Drawing on its own internal sources, the satellite suspended from the manipulator arm of the space shuttle slowly comes out of the cargo bay.

The satellite is poised at the end of the extension of the mechanical arm, suspended by a steel claw out in space, floating with the shuttle in a hurtling but seemingly motionless path above the earth. The woman waits for a sequence of events to be relayed through the mindless electronics of the orbiter, and for an equally mindless sequence of electronic thought to happen aboard the satellite.

The satellite unfolds its solar panels. It is no longer a steel, aluminum, platinum, gold foil chrysalis in the cargo bay. It slowly moves away from

105

Living aboard the orbiter will not be luxurious for the seven-person crew, but it will be the most sophisticated ever in space. (Courtesy NASA)

the clamps of the mechanical arm. The shuttle maneuvers back to its original orbit. The satellite begins telemetry to earth.

Among the many things which separate the space shuttle from previous space vehicles—size, the ability to land on a runway, normal earth atmosphere in the cabin, reusability—nowhere is it more different than in crew selection. The space shuttle is designed to carry what are essentially passengers—nonastronauts.

While America's largest previous spacecraft, the Apollo, could hold only three astronauts (later modified to hold five as a rescue vehicle for *Skylab* crews), the space shuttle is designed for a standard crew of seven and will hold ten people in an emergency. Of the seven standard crew, only three will necessarily be NASA employees and only two will be what would normally be thought of as astronauts.

If the planned five hundred flights of the space shuttle all take place during the next ten to fifteen years, as many as two thousand ordinary people will have been into space by the last decade of this century.

Though some will be repeat travelers, the trip will be a once only and a first time for a great many others. Before the space shuttle, only twelve men had landed on the moon, and there were fewer than fifty Americans in space during the entire space program. (Actually there have been 71 trips into space for 43 astronauts on 31 manned flights. Four American astronauts, for example, have flown four times.) With space shuttle in full operation, toward the early 1990s, one American in every 200,000, on the average, will have been into space. On the same average basis, a city the size of Los Angeles would be able to boast at least 35 space travelers.

Not only will the space shuttle be taking more people into space than was even talked about during the early days of space exploration, but women will be going into space for the first time in the American space program (there has been one Russian woman cosmonaut), as will minorities and passengers from other countries who will be taking care of shuttle payloads belonging to them.

There are three classifications of shuttle crew: commanders, copilots, and mission specialists. A fourth category, the payload specialists, are what are generally referred to as space shuttle passengers, since they are trained for a short period and deal only with one special or particular payload, and are not necessarily NASA employees or astronauts.

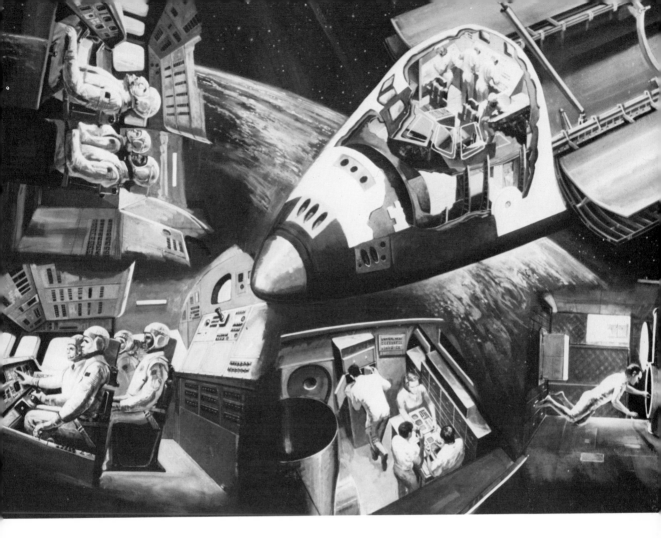

After the space shuttle becomes operational, there will almost certainly be VIP flights in which men and women will go to space basically just for the trip and for public relations. If Wernher von Braun were still alive, he would be a prime candidate for a VIP trip. Walter Cronkite has been frequently mentioned in this context as have several well-known science figures such as Carl Sagan and Arthur C. Clarke.

Most of the men and women who will crew the space shuttle will be new to space. The total of available astronauts from all previous manned programs is only twenty-seven, which includes the scientist-astronauts left over and available from the Apollo missions and the men who joined the astronaut corps when Air Force's Manned Orbiting Laboratory pilot program was disbanded.

The original astronaut training was comparable to the ecstasies of torture introduced by the Holy Inquisition: centrifuge training to twenty times the force of

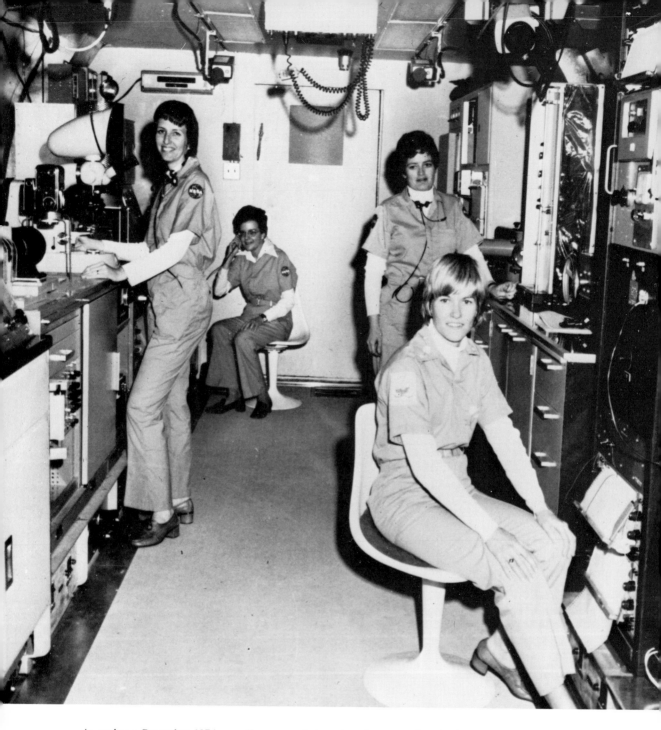

As early as December 1974, an all-woman mission aboard was conducted aboard the general purpose laboratory at the Marshall Space Flight Center. The crew conducted materials processing experiments of the type which will be carried aboard spacelab. (Courtesy Marshall Space Flight Center)

gravity—20g—until the victim's capillaries were bursting and their teeth were the only recognizable feature of the face. Shut inside ovens at 140 degrees Fahrenheit for up to two hours and then thrown into an ice pack to see how they reacted (It is hoped that the early astronauts verbally indicated their distaste for the flight examiners). Taken into hostile jungle and told to eat whatever was available, flown to burning deserts to survive as they could. Worst, shut into cramped spacecraft environments (A capsule, the astronauts were forever trying to tell the press and public, is for swallowing. A *spacecraft* is what we fly) which would drive a normal man mad in a few minutes, screaming and yelling to be released from the uncomfortable aluminum maiden.

They were in truth almost supermen, trained for eventualities beyond belief; exploring an unknown territory; able to cope with any stress, any pressure; watched like laboratory rats and treated, most of the time, with about the same human courtesy as the rats got. And they are nearly all gone.

Only one of the original Project Mercury astronauts is still on flight status, Donald K. "Deke" Slayton, originally in the Air Force, now a civilian attached to NASA. Deke Slayton waited the longest to get into space (a minor heart problem kept him grounded all during Gemini and Apollo), finally flying once as the docking module pilot on the prime crew for the Apollo-Soyuz Test Project. Of the second group of astronauts, selected in 1962, only John Young is still on flight status. He has flown in space four times and landed on the moon.

From the astronaut corps picked between 1963 and 1965, there are only four men still on flight status: Alan Bean, Owen Garriott, Edward Gibson, and Joseph Kerwin. The last three are all scientists, Kerwin is a medical doctor.

Obviously, for the space shuttle many more astronauts and other crew members are needed, and it was for this reason that in 1978 NASA selected—from almost eight thousand candidates—thirty-five to form the nucleus of a new astronaut corps which would see the space shuttle into its first years of operation.

The requirements, relative to the early days of the space program, were very relaxed. The space shuttle's launch and re-entry forces were so much less than previous spacecraft that extensive testing of ability to survive gigantic g forces was no longer necessary. Other factors learned from the special corps of scientist-astronauts showed that the rigid requirements of having been a military test pilot were also no longer necessary, although highly desirable for commanders of the shuttle.

Even the requirements of 1978 may be lowered considerably as new groups are chosen for the operational phase of the space shuttle in the 1980s.

One of the first changes in the NASA requirements was an invitation to women and minorities to join the astronaut corps. Despite rumors that NASA advertised in women's magazines such as *McCall's*, and sent recruiters into depressed urban areas, the women and minority applicants who arrived at NASA did so in the same way as all previous groups: via bulletins seen at their university or distributed in their corporation's newsletter, or announcements of opportunity in any one of a dozen scientific and technical magazines and journals.

Toward the end of the application period, NASA did send out Michelle Nichols, who played the communications officer Lieutenant Uhura in the television series "Star Trek", but that apparently had little effect on the recruiting results except to give Trek fans more ammunition for their belief that NASA was about to accept them as spiritual partners in the exploration of space. Actually there had been talk for years of having women in the astronaut corps and it was only the abrupt closing of the Apollo Program with *Apollo 17* (more than twenty flights had originally been scheduled) which precluded women getting on the few later missions flown; Skylab and Apollo-Soyuz. Had there been women military test pilots available at the time it is speculative, but probable, that women would have gone into space with Apollo.

The second major change in astronaut selection was the lifting of the age limitation. For space shuttle, there is *no age limitation* for commanders, pilots, or other crew members.

The third major change was because of the space shuttle's unique crew possibilities: there would be at least two sets of criteria for flight into space: one for commanders and pilots, and one for mission specialists and payload specialists. Passengers, or VIP mission crew, would fall loosely into the second category.

For commanders and pilots aboard the space shuttle, the first requirement is a bachelor's degree in one of the physical sciences, mathematics, or engineering. The second requirement is for two thousand hours or better in jet aircraft, with one thousand hours as a first pilot. The preference is for those whose jet hours are in high performance aircraft. Selection is not limited to them, but military test pilots still have an advantage. To pass the physical for shuttle flight, a prospective astronaut must go through the equivalent of a military Class I series of tests.

For those who have never undergone a Class I physical in the military, it is approximately equivalent to being accepted to an Olympic team for people with the emotional stability of the average granite rock. Its requirements are not as exacting rigid as for the early astronauts, but are sufficiently rigid to keep out all but the best qualified pilots from military and civilian sources. Commanders and pilots must have vision of 20/50 uncorrected or 20/20 if corrected.

The mission specialists aboard the space shuttle must have a bachelor's degree in engineering, mathematics, or one of the physical sciences. There are no basic restrictions for the job, but the physical is definitive. They need not be pilots nor necessarily have an aircraft pilot's license, although it is expected that quite a

Fred W. Haise Jr., commander of the first crew which flew the space shuttle (during the approach and landing test program in 1977) in the cockpit of the orbiter aeroflight simulator at Johnson Space Center. The simulator can duplicate almost any flight environment of the orbiter. (Courtesy NASA)

few will be good pilots of small and medium sized aircraft. They must have good vision, almost as good as a pilot's.

For a payload specialist who goes on a specific mission, the requirements are down to the level which many people could sensibly meet. It is desirable, but not mandatory, that the crew member have a master's degree in one of the physical sciences or mathematics. A NASA ability test must be passed, and uncorrected vision of 20/100 or 20/20 with correction is required. The physical is a Class II, similar to an ordinary military physical.

For true passengers on the VIP flights and similar NASA ventures, the only requirements will be that the person have a reason for going into space (its worth

In 1976, one-week missions were duplicated in the spacelab simulator at Johnson Space Center. (Courtesy Johnson Space Center and NASA)

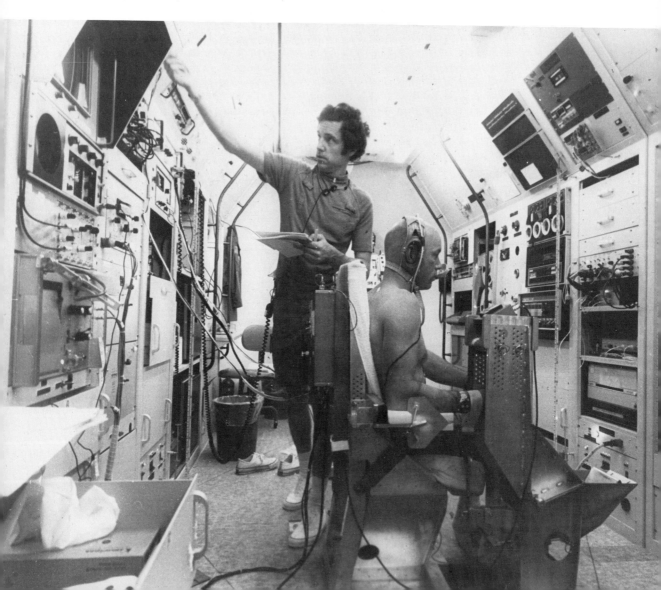

*Future astronauts train on the shuttle orbiter simulator at Johnson Space Center. This simulates the primary flight control station of the **Enterprise**. (Courtesy Johnson Space Center and NASA)*

decided, obviously, by NASA) and be able to pass the Class II physical. In simple terms, that could be just about anyone.

All applicants for crew will be tested by psychologists. Physical, neurological, muscular, skeletal, and eye tests will be made, and EKG baselines taken, as well as electro-encephlograph and treadmill testing. All applicants, especially for mission specialists and payload specialists, must pass a test in the rescue sphere.

The duties of the space shuttle crew are mainly separated by phases of the mission. The commander and pilot are responsible for the take-off, orbital insertion, deorbit, and landing. Mission specialists will be responsible for coordinating space shuttle operations which deal with space walks and payload handling and maintenance, and they will work with the scientific experiments when there is time or they are needed in a particular instance. Almost all of them will be NASA employees.

Payload specialists, on the other hand, are employees of the agency or agencies who are paying for the orbiter's cargo on a particular mission. If a payload is a completely NASA project, then all of the crew will probably be NASA employees. Department of Defense payloads will have military specialists. But other customers—Western Union, other countries launching communications satellites, Spacelab—will send their own employees as the payload specialists.

The payload specialists will be trained by NASA for a short period—just enough to enable them to cope with space once—and then they will go into near-earth orbit with their piece of equipment or set of experiments. Some payload specialists will be more highly trained and will be responsible for on-orbit maneuvering and payload deployment.

The crew of Spacelab 1 is a good example of the "new astronaut" concept. It will be flown aboard a space shuttle in the early 1980s. Five European and American scientists were selected in July 1978 to operate the experiments aboard the flight. They will be payload specialists for spacelab.

These five were chosen by the scientists who had designed the experiments that will go into space on the mission. After spending three months training to operate the scientific instruments, they began NASA training in January 1979. NASA training will last for almost three months in nine cities and two NASA centers. Nearly two weeks will be spent at the manned spacecraft center in Houston, Texas.

When the spacelab mission lifts off, only two will actually fly in space; the other three will operate ground support equipment. In flight aboard the spacelab, the two payload specialists finally selected will have to operate more than forty scientific instruments. Spacelab 1 will investigate space plasma physics, materials processing, solar physics, and medicine on its seven-day flight.

In the same spacelab mission will be two mission specialists. In this case, no new astronauts were chosen to tend the orbiter and its systems. The Spacelab 1 mission specialists are Dr. Owen K. Garriott and Dr. Robert A. Parker. Garriott was one of the original Apollo scientist-astronauts selected in June 1965. He has advanced degrees in electrical engineering, has been an assistant director for space sciences at the Johnson Space Center, and flew on the second manned *Skylab* mission for 56 days in space.

The other mission specialist, Robert Parker, was a mission scientist and spacecraft communicator during Apollo and Skylab. He has never flown in space. Parker was selected in the August 1967 group of Apollo astronaut-scientists, and has a doctorate in astronomy.

The commander and pilot for Spacelab 1 will be taken from the older astronaut corps of highly trained military pilots and both will have had one or more missions in space.

Since the astronaut corps as a whole is not as large as it needs to be, during early space shuttle flights astronauts who have flown one or more missions will be

Shuttle astronaut training includes extensive time in the buoyancy tank at Marshall Space Flight Center, which can duplicate the feeling of weightlessness and the general working conditions in space. Full scale mockups of shuttle sections can be placed in the tank. (Courtesy NASA)

paired with those who have not flown or who have been on only one mission. The same will be true of the mission specialists and if possible, the payload specialists.

For the example of the Spacelab 1 flight, the crew of six will be divided between two operational shifts. Each shift will have one of each category on duty: pilot, mission specialist, and payload specialist. Of the five payload specialists selected for training, one is German, one Dutch, and one Swiss, the remainder are American. Four are from universities and one is from the European Space Technology Center.

The new NASA astronaut corps is much more diverse than the old. Of the thirty-five candidates selected in 1978 (none of which were selected for payload specialists, but will fill the commander, pilot, and mission specialist categories),

21 are military officers and 14 are civilians. Six are women and four are minorities. All of the women are candidates for mission specialist. Since none of the women have flown in space previously and they will not finish their preliminary training until after the space shuttle's first orbital flights, the first woman into space may be still some years off, perhaps about 1981.

Pilot training is similar to that of Apollo. The pilot candidates use a shuttle orbiter trainer which duplicates some of the controls of the real vehicle. The trainers can set up flight possibilities and test the candidates' reactions to them. For the shuttle training, only the primary flight control station of the shuttle's flight deck is duplicated.

For more specific and more complicated training, the pilot-astronauts use the Orbiter Aeroflight Simulator at the Johnson Space Center, Houston, Texas. The flight simulator is a complete unit designed to support both crew and flight controller training. The effect is as near as possible to an actual flight, complete with potential emergencies. The simulator has been responsible for the ultimate trainee exchange: "How'd it go today?" "Not bad, but I crashed the shuttle."

By the time a pilot has passed the simulator phase of the astronaut training, he or she will be able to satisfy all requirements for approach and landing of the orbiter, launches, and aborts. For additional on the spot training, the orbiter's actual cockpit is used. The final phase is a flight with an experienced astronaut-pilot as the commander.

All of the crew members train in the water, sometimes with scuba gear, in the neutral bouyancy simulator at one of the NASA centers. Being completely underwater closely duplicates some of the problems involved in working in space and experiencing weightlessness.

During some of the training in the water tanks, the crew wears the space shuttle extravehicular activity suit. When women joined the astronaut group, the company which manufactured the suits, Hamilton Standard in Connecticut, had to make some changes. Some extravehicular activity suits are now made in extrasmall, and parts of the suits have been redesigned to make them easier for women to put on.

In the deep water tanks, it is possible to make the trainees feel as if they are actually in space. They are in suits and feel nearly weightless. Full scale portions of the orbiter's cargo bay areas can be lowered into the tank. A trip down in the tank with a suit on, waving about uncertainly, losing motion, then finally clawing through the water toward a section of the cargo bay using cables and attachments, with "Houston" whispering in the earphones and simulated conversation with the orbiter crew, is about as close to a shuttle flight as anyone can get. All of the sensations are there. If they painted the tank all around with stars and constellations and a giant earth stretching over half the space, it would be frighteningly real.

An engineer wears the shuttle space suit in tests at Johnson Space Center. The space suit will be worn by pilots, copilots, and mission specialists for extravehicular activities and emergencies. (Courtesy NASA)

Other training takes place with the rescue sphere. The sphere is the primary unit for emergencies where an orbiter is crippled for any reason. Another orbiter would be flown into space with a minimal crew and its orbit matched with the orbit of the crippled ship. If the crippled ship was stopped in space and not spinning, tumbling, or moving in other strange ways, they might be tethered together. Then the long manipulator arm of the rescue shuttle could be used to lift each rescue sphere complete with crewmember out of the crippled ship and deliver it to the second orbiter. Other crew would transfer by extravehicular activity suit.

If a crippled orbiter were tumbling and out of control, tethering would be impossible. In that case, the crew members with extravehicular suits—commander, pilot, mission specialist—would use propulsive backpacks to ferry the rescue spheres over to the waiting rescue orbiter. The so-called "beach balls" or rescue spheres can support life for 90 minutes. The extravehicular activity suits are habitable for hours.

Apart from prospective VIP missions in which some crew would be strictly passengers in the ordinary sense, there are no present plans to turn the space shuttle into a tourist ship.

But if a space shuttle can carry a spacelab, it can carry other things of a similar size in the cargo bay. There may come a time in the later operational phases of the space shuttle when a sightseeing module may be added to the cargo bay. About the size of a Greyhound bus, the tourist module would have windows like a railway dome car made of the same materials as the orbiter windows.

The tourist module would be supplied with light, heat, and oxygen by the orbiter and would be put into near-earth orbit for a few hours or even the better part of a day. On a one-day mission, the tourists would be treated to the sights of the earth's curve, the sun and moon from orbit, and perhaps a tour of a space construction base and a drift by a space telescope in orbit.

In their space bus they would be as safe as any other payload. For emergencies, the shuttle would carry an extra couple of loads of rescue spheres, and a backup rescue shuttle complete with another tourist module would be ready on the ground.

How many could go into space as tourists on one shuttle flight? It would depend on how much space was taken up with life support, rescue spheres, and sanitary facilities. The inside of a tourist module might resemble the first class cabin of a 747 jet, complete with stewards or stewardesses and a small galley for meals. As many as thirty people might be comfortably fitted into a bus which would fit the space shuttle's cargo bay.

The cost? Very high. There is no exact way to even guess, since tourist modules haven't been designed. But it might be as low as $10,000 per person at some point in the shuttle's lifetime. And that's not much more expensive than a round-the-world trip today.

During the period of the Saturn-Apollo missions we were pilgrims in space, ranging from home in search of knowledge. Now we will become shepherds tending our technological flocks, but like the shepherds of old, we will keep our eyes fixed on the heavens.

President Jimmy Carter, 1978

7

THE WORLD OF THE SPACE SHUTTLE

The space shuttle carrying earth satellites has long since landed. The crew have gone back to their old jobs or to Johnson Space Center, or have remained near the shuttle launch site at the Cape. Some will go into space on the next shuttle flight; some will wait out their duty time in planning and briefing sessions; some will never go to space again and they will never forget the experience.

Rising now into low orbit is another shuttle, a special one. Its cargo is another rocket. Connected to the tip of the rocket is a complex structure of gears, wires, cameras, scoops, and high-gain and low-gain antennas—a planetary exploration spacecraft.

Shuttle cargo bay doors open, and the long rocket is moved outside and away from the space shuttle. The crew will next see it in a portion of the television picture broadcast from the surface of Titan, the giant moon of Saturn. Voyager, the unmanned probe, has already spent its days in photographing the surface of the moon, testing the atmosphere. Pioneer, two years before that had sent back some sketchy photographs of the cold silent surface under an atmosphere where complex molecules might have chained and lived.

Now it is time for mankind to see the surface of Saturn's moon, dig and bite at the rocks and ice, mechanically breathe poisonous air from a great distance, and read electrons to guess the temperature and consistency of a far and distant land.

The shuttle is ready; the mechanical arrangements have been made. The shuttle moves away from the rocket and spacecraft, and the engines fire. In a few minutes the payload is lost from sight among the stars. It will wander for four years before the gravitation of Saturn pulls and pulls and distorts its orbit and the lander insulated in an aeroshell breaks away to find the surface it has journeyed so long to meet.

A space shuttle payload is most often thought of as some spidery aluminum device of high technology, costing a few million dollars or more, or large single pieces of equipment such as spacelab, Space Telescope, and large interplanetary exploration vehicles.

This is not always the case. Shuttle has plenty of capacity for the large pieces and single mission loads but it also has room for the experiments of medium sized companies, small firms, and even individuals.

While the largest users of the space shuttle's cargo bay such as Western Union, the Department of Defense, and NASA must deposit $100,000 each with an average mission cost as high as $20 million, the small users can deposit as little as $500.

NASA's getaway special for small experiments and cargo costs as low as $3000 total for 1.5 cubic feet (.042 cubic meters), and as little as $10,000 to orbit a load as large as 5 cubic feet (.14 cubic meters). For these bargain basement prices, NASA allows a minimum weight for a getaway special of 60 pounds (27.21 kilograms) and up to 200 pounds (90.72 kilograms) for the larger package.

NASA has four general criteria in accepting applications for the getaway special: the experiments must be of a broad scientific nature; they must be safe, i.e., nonexplosive; and they must be self-contained. They must also meet a nebulous qualification of good taste.

The NASA definition of good taste has been used to eliminate what is *not* in good taste. So far, with space shuttle flights through 1982 nearly sold out, there have been several interesting rejections for cargo. A retailer in Dallas wanted to send his gerbils into orbit so he could get a better price for them at the Neiman-Marcus department stores. He was refused cargo space on the shuttle. So was a New York doctor who wanted to orbit a small satellite as a wedding gift for his son. And there will be no stamped envelopes, medallions, sculptures, paintings, or coins sent aloft aboard the shuttle.

Among the items which have been approved for flight, at least to the point of

PUT YOUR PRODUCT OR EXPERIMENT INTO ORBIT

ON THE

SPACE SHUTTLE

SPACE AVAILABLE NOW!

From
SURGE CORPORATION
Box 922   Palestine, Texas 75801

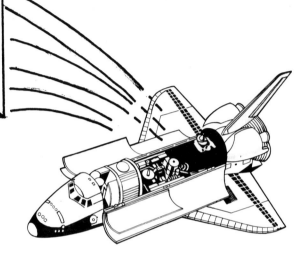

Now you can gain a competitive edge in the space technology market. You can prove your product in the most severe environment man has ever challenged. You can develop original processing techniques; space qualify your product for broader markets; evaluate equipment and experiments. Write or call today for complete details. Prices start at less than $200 complete.

PHONE AC 214-729-3201

*This advertisement appeared in 1978 in **Science News**. The space shuttle is not only for use by large corporations; schools, individuals, and small companies will use the shuttle for experiments in near-earth orbit. (Courtesy **Science News**)*

accepting the deposit, is a package from Steven Spielberg. Spielberg's camera package will be used to film the earth and moon from the space shuttle to get "special" special effects for a sequel to CE3K. For those who have not been infected with the use of acronyms, CE3K stands for *Close Encounters of the Third Kind,* a shortened form used by the movie industry and fans alike.

Volkswagen has reserved shuttle cargo space to test manufacturing in weightlessness. VW is interested in making the "perfect ball bearing." Johnson & Johnson will carry out experiments in separating the chemical constituents in human blood. The *Yomiuri Shimbun,* Japan's largest newspaper, has reserved space for the best scientific experiment designed by a young Japanese.

The list is becoming large as the space shuttle gets closer to its operational phase. By mid-1978, 240 orders for small spaces had been received by NASA. The total value of the cargo was in excess of $520 million, a perfect argument against those who think the shuttle is too expensive and should never have been built. Early indications all point to a very profitable space shuttle operation for NASA, instead of the high exploration costs of the Apollo program.

Beginning with the operational phase of the space shuttle in 1980, the orders already received would fill up more than twenty flights.

A Boy Scout troop will be sending an experiment into orbit. A high school student, future undergraduate at Utah State University, will orbit an experiment dealing with bacterial growth in zero-g, in this case, the bacteria which cause tooth decay. San Diego Community College, Prairie View (Texas) A&M College, and Weber State College in Ogden, Utah, are among the educational institutions that will take advantage of the small user class of cargo space aboard the shuttle in the early 1980s.

The use of the getaway specials is not limited to citizens of the United States; the space shuttle is, at least in a cargo sense, a very international affair. Egypt has reserved the space for four small self-contained payloads. The Egyptian payloads will be in the 200 pound (90.72 kilograms) class and will require no additional space shuttle services such as electrical power or deployment in space. Although Egypt's reservation in 1978 is the first foreign educational use of cargo space aboard the shuttle, it will be followed by hundreds more from a score of different countries. One of the Egyptian payloads will feature an experiment designed by Egyptian students who will compete in a nationwide contest by submitting proposals.

One of the main payloads aboard the shuttle will be a wide variety of earth satellites. Since the shuttle can carry a multiple payload of earth satellites, a mission might be a mixed bag of hardware from different countries with an American satellite thrown in. India has purchased space aboard a shuttle satellite mission for the launch of the Indian National Satellite System-1.

Scheduled for launch in the first quarter of 1981, the Indian satellite will be a first of its kind. Remaining always over one spot above the earth, it will provide domestic public telecommunications, direct television broadcasting, and weather services for India. India was the first country to sign up for a shuttle launch in which NASA will be paid for the launch services.

Space shuttle will be the key to the opening of near earth space to extensive satellite networks for all countries. In a demonstration of what the future satellite networks can do, in 1978 NASA and the Canadian Department of Communications broadcasted the United States Conference on Technical Cooperation Among Developing Countries held in Buenos Aires, Argentina, with the Communications Technology Satellite. Cooperating with NASA in this demonstration were the

An earth services satellite is released in orbit by the space shuttle. Shuttle can carry multiple satellites of all kinds into low-earth orbit. (Courtesy Rockwell International Space Division)

Communications Satellite Corporation, COMSAT, and an Argentina telephone company called ENTEL.

In this "shape of things to come" demonstration, the picture and voice of the people attending the conference were sent to a portable NASA terminal at the United Nations in New York. There the transmissions from Buenos Aires were translated into the five official U.N. languages and sent back to Buenos Aires via the Communications Technology Satellite and a portable COMSAT terminal in the city. The U.N. delegates could select any of the five languages they wished.

Another component of the same system sent copies of all the speeches and other documents from Buenos Aires to New York. They were translated and then sent back the next day, ready for reading. The CTS is one of the most powerful communications satellites now in space; it is possible to send and receive color television, voice, and data simultaneously in *both* directions between two earth terminals.

What this will mean eventually, thanks to the shuttle's ability to deliver multiple payloads of earth satellites of all kinds into low earth orbit, is greatly increased use of satellite communications by all countries of the world. In the not too distant future, the United Nations sessions, for example, will be conducted entirely by satellite with the delegates residing in their own countries instead of in New York. They will speak, listen, receive messages, documents, and conduct their United Nations business by color television in their own language and in their own time zone.

Not only will the shuttle payloads be communications satellites of many countries, but also more sophisticated weather satellites, developed from the United States' Tiros series, which will keep track of weather for many countries. Smaller countries will share the expenses of orbiting a complex weather satellite; larger countries may have several in space.

Among the scientific satellites which will form other shuttle payloads will be advanced versions of the International Sun Earth Explorer. These exploration satellites will be sent, like their predecessors, to the sun-earth libration point, at which the gravitational pull of the sun just balances that of the earth-moon system. Although to an earth observor they would appear to be orbiting the sun, in fact they would trace a halo above the earth, completing a revolution every six months. The libration point is about one million miles (1.6 million kilometers) from the earth.

From this vantage point, an International Sun Earth Explorer will measure the solar wind constantly emitted by the sun, sun spots, and solar flares. The goal is to better understand how the sun influences the earth's fluctuating near-space environment and its atmosphere: weather and climate, energy production, and ozone depletion. With shuttle-launched ISEE spacecraft, the earth will have an early warning system for activity on the sun. Had such information from satellites been available a few years ago, *Skylab* might have been placed in a higher orbit and would now not be in danger of falling to earth because of orbital decay due to solar activity.

The scientific satellites which will be launched by the space shuttle will also be international. A solar activity satellite like the ISEE might be worked on by as many as one hundred investigators representing more than thirty universities and ten nations. In fact, large parts of the satellite would be built in other countries and shipped to the United States for final assembly and launch at Cape Canaveral by shuttle.

Among the other scientific satellites which may be sent aboard shuttle are improved versions of the International Ultraviolet Explorer. Built by Americans, British, and Europeans, the first Ultraviolet Explorer satellite could see into the core of globular clusters of stars in our galaxy which were more than 15,000 light years away. That incredible distance would be the equivalent of fifteen thousand times 6 trillion miles—the number of zeros would fill nearly a full line of this page. A globular cluster is a miniature galaxy of tens of thousands of tightly packed stars. Our galaxy, the Milky Way, has about 150 globular clusters scattered, apparently at random. Most are regarded by scientists as nearly as old as the galaxy itself.

The International Ultraviolet Explorer also first discovered the possibility of a massive black hole at the center of some groups of stars in our galaxy. A black hole is believed to be the final stage in the collapse of a massive dying star. The collapsed star's material is so densely packed that even light waves are unable to escape the gravity.

Although other governments will be customers of the space shuttle for their satellites, a large portion of the shuttle's payload capacity will be taken up by launches for NASA, the Department of Defense, and private corporations in the U.S. The Pentagon has reserved space aboard the shuttle for early-warning watchdog satellites; NASA has plans for an extensive array of scientific exploration satellites; and RCA, COMSAT, and Western Union will be launching large payloads into earth orbit. Some of these large payloads will be fitted into the cargo bay with small payloads from schools, colleges, and individuals. It was not by accident that someone named the shuttle "the world's fastest freight train."

In addition to the various communications satellites and weather satellites which will go to space in the early years of the space shuttle's operational phase, one of the first of the NASA payloads will be the long duration exposure facility. This is a basic research project implemented by NASA but expected to be used by hundreds of individuals and companies, and several governments. It is called a free-flying structure because it will be detached from the space shuttle once in orbit and left there for extended period of time.

Aboard can be an experiment which requires an exposure to the conditions of space beyond the standard space shuttle mission of seven days or the extended mission of thirty days. The facility is modular, convertible to a wide variety of

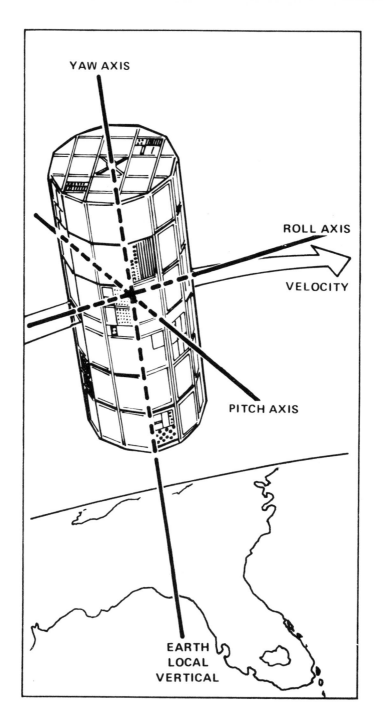

The long duration exposure facility can support a wide variety of experiments that must remain in space between shuttle flights. (Courtesy NASA)

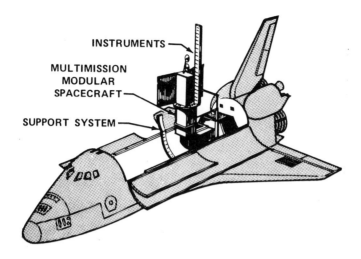

The multimission modular spacecraft is reuseable and can support an array of operational and research instruments. (Courtesy NASA)

uses, reuseable, and costs very little. It can be loaded with experiments which are paid for by a dozen companies and organizations, flown to near-earth orbit aboard the shuttle, and recovered when the experiments are finished by a later shuttle flight from two months to two years later. It can then be stripped of the old experiments, fitted with a new set, and flown again.

The first flight of the long duration exposure facility as a shuttle payload will go into space carrying a microdot containing hundreds of handwritten messages which were recorded by visitors at the NASA Langley Research Center during the twentieth anniversary of the first American spacecraft, the *Pioneer-1,* launched October 11, 1958. The messages will stay in orbit along with the facility from six months to a year. A microdot the size of a postage stamp could contain as many as ten thousand signatures. Among the messages which will orbit the earth are "Bring back peace to all mankind," "God be with us," "one nation," and simply "shalom."

Another plug-in NASA payload for the space shuttle will be the multimission modular spacecraft. This is a basic core satellite—usually called a bus—to which can be added experiments and attachments for a series of different tasks and investigations. The multimission modular spacecraft may become extremely useful in the early years of the space shuttle because it is cheap and much easier to design than a one-off satellite built for a specific purpose.

The modular spacecraft might go up in one shuttle flight with a selection of communications equipment which is being tried out, while on another mission it might contain only weather gear to test improvements in weather satellite technology. Like so many of the other space shuttle designs, its first virtue is reuseability which lowers cost. The second is the versatility which allows extensive testing of equipment before a company or a government invests in a single satellite for a specific purpose.

Using the space shuttle for launching, the world of the 1980s and 1990s will be ringed by systems of useful satellites: communications, weather, navigation on a global scale; Landsat type satellites to monitor earth resources, minerals, and pollution; and Seasats to investigate the oceans. Military satellites will be so accurate that details smaller than a car license plate can be easily photographed on the earth's surface. In the not so distant future, personal communications satellites may be sent up by the space shuttle or built in space from materials sent up by shuttle. These personal communications satellites will allow an individual to dial a phone call on a Dick Tracy wrist communicator to any place on the earth and receive calls from all over the globe.

Since shuttle is reuseable and multiple trips to space are easily made, it will be possible in the future to build complex and heavy spacecraft for planetary exploration by unmanned vehicles. Previous planetary craft had to be relatively small because they had to be sent to the target in one rocket on one flight from Cape Canaveral. Built in space from materials sent up during three or four shuttle flights, future planetary craft will be heavy with experiments and many will contain a lander craft to travel down to the surface of a planet or moon.

Already designed and scheduled for launch aboard the shuttle is the *Galileo* Orbiter-Probe, formerly named the Jupiter Orbiter Probe. It will be launched by shuttle in 1982 or 1983 for rendezvous with Jupiter in 1984–1985. Future missions to Mars with heavier Viking spacecraft are possible, as well as sample-return missions to the nearby planets of Mars and Venus. In the more distant future, landers may visit Ganymede, a moon of Jupiter, and Titan, a moon of Saturn. They may rendezvous with asteroids out in the asteroid belt. Ultimately, large planetary exploration spacecraft launched by shuttle may travel the entire solar system investigating each planet and reporting back to earth. Probes may eventually go to the outer reaches of the sun's influence beyond Pluto.

Space shuttle will also be the launch vehicle for the large single-mission payloads such as spacelab and the Space Telescope which will be discussed in later chapters. And in the future, shuttle may take materials, structural beams, habitability modules, and power station parts into space to begin the building of space stations and construction bases for the coming industrialization of near space and the commercial use of the space frontier.

To do all this, the shuttle needs accessories and it has them. This is the reason why the whole project is called the Space Transportation System, instead of just the space shuttle. Like so many other major pieces of machinery—such as

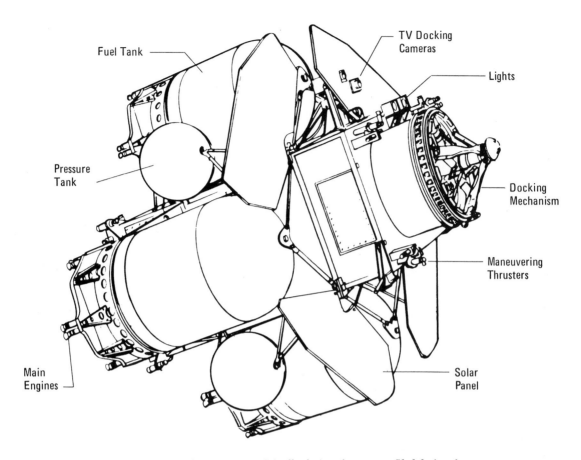

Fuel Tank

TV Docking
Cameras

Lights

Pressure
Tank

Docking
Mechanism

Maneuvering
Thrusters

Main
Engines

Solar
Panel

*The teleoperator retrieval system was originally designed to rescue **Skylab**, but that program was disbanded late in 1978. It can be used with the shuttle for many future mission possibilities. (Courtesy Martin Marietta Aerospace)*

automobiles—the shuttle has a much wider range of abilities with the expensive extras.

One of these accessories is the Teleoperator Retrieval System. It is a remote controlled device designed to be taken into space in the shuttle cargo bay. Once in orbit, the TRS will inspect, dock with, retrieve, and maneuver spacecraft, satellites, or anything else which might need moving from one place to another. It is a version of the once much discussed concept of a space tug.

The original retrieval system contract scheduled delivery at the Cape by late August 1979, in time for the first of the orbital test flights. In addition to its duties as a space tug, it can be used to carry some scientific payloads—serving as a bus for these items.

It can be used, instead of the shuttle, for sending satellites to earth-orbit higher than the shuttle. It can retrieve satellites from very high orbits and return them to the shuttle or to a station in space where they can be repaired. Manipulator arms or steerable high-gain antennas for long-distance communication can be added easily, quickly, and most importantly, cheaply. The system can also be used for emergency repairs.

One major advantage of the TRS is that it can go up to earth-orbit aboard shuttle, be released and guided to a distant and large payload, retrieve the payload, and bring it back to the shuttle. The large payload can then be taken down to the earth for repair. Meanwhile, the teleoperator retrieval system is placed in a parking orbit high above the earth, waiting for a shuttle to take it down again, or waiting in space for another shuttle flight to use its services. The space shuttle, of course, is also capable of retrieving payloads with its manipulator arm, so the teleoperator retrieval system will not always be used when a satellite is serviced, repaired, or retrieved. If large structures are finally built in space, it will probably be the TRS which will open the way for the beginning of their assembly.

The teleoperator retrieval system has its own propulsion system, and it can be supplemented with extra propellant kits at any time. It has its own on-board computer for guidance and attitude control, or it can be controlled by a shuttle crew member using the control boards at the orbiter payload station. The basic TRS is box-like, 4 by 4 by 6.7 feet (1.2 by 1.2 by 2.0 meters). It has a docking system on the forward end of the core structure, complete with two television cameras.

One of the first uses for the TRS was to be the rescue of the ailing *Skylab*. The retrieval system was supposed to boost *Skylab* into a higher orbit where it would be safe and useable in the future. It was thought there was a possibility that the orbit would decay in 1978, allowing *Skylab* to drop down to earth. If it comes down by itself, *Skylab* will, of course, partly burn up in the atmosphere. Since *Skylab* was still a very useable piece of equipment, though admittedly a little obsolete, it was suggested that it be saved, perhaps to become the core of a space station for the first few years of space shuttle's operational lifetime.

Information released by NASA very late in December 1978 indicates that the *Skylab* orbit is decaying much more rapidly than was thought. *Skylab's* orbital lifetime was underestimated, mostly through incomplete information on the effect of solar activity on the orbit. It is possible that *Skylab* may fall to earth as early as May 1979. Delays in the TRS and the space shuttle system will not allow a rescue of *Skylab* even by the first shuttle orbital test flight in the fall of 1979.

The interim upper stage is another piece of space shuttle equipment which will fit in the cargo bay. IUS is an extremely versatile, multistage, solid-fuel rocket designed to put shuttle payloads into orbits which shuttle or shuttle/TRS

The manipulator arm places a satellite and the interim upper stage into position for launch into a high orbit beyond the shuttle's capabilities. (Courtesy NASA)

can't reach, or into interplanetary orbits. It will also be used as an upper stage to the United States Air Force's Titan III and a new military expendable launch vehicle called the Titan 34-D.

In the original order for the interim upper stage, there were plans for nine of the vehicles, four for the Department of Defense and five for NASA. The four Department of Defense interim upper stages will all be Titan-launched from the earth's surface, using the IUS as a second stage. Thus, it will basically replace the currently used Centaur upper stage, which launched the interplanetary missions of *Viking* to Mars, and *Voyager* to Jupiter and Saturn. These launches will, of course, be used for military purposes.

Of the remaining five interim upper stage vehicles, one will be used to launch the Jupiter *Galileo* Probe to orbit the giant planet and send probes deep into its atmosphere. *Galileo* will be the first interplanetary mission launched with the space shuttle.

Like most of the space shuttle equipment, the IUS has been carefully designed to be as versatile as possible, so that over the next twenty years of the

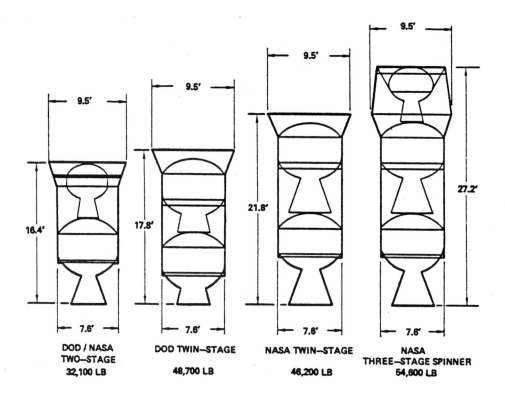

Four configurations of the interim upper stage. (Courtesy NASA)

An artist's conception of the interim upper stage in orbit near the space shuttle, which is preparing to launch another IUS. (Courtesy Boeing Aerospace Company)

United States space program, it can perform relatively inexpensively most of the tasks set for it by the scientific, commercial, and military users.

It is a basic two-stage vehicle, using components which have already proved themselves in past space programs. But it is designed for a building-block approach and can come in four sizes. There are, in reality, four types of IUS: the Department of Defense/NASA twin-stage basic; the Department of Defense

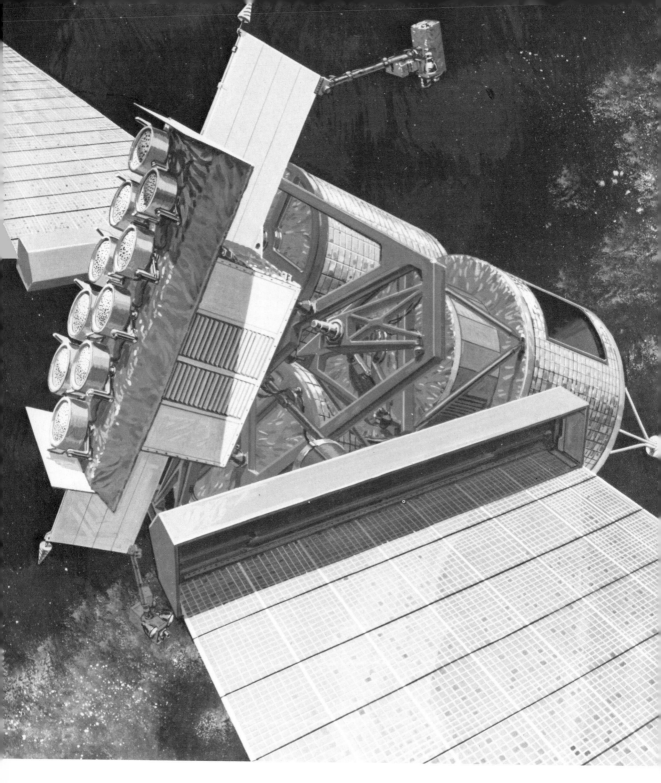

The solar electric propulsion system is proposed by NASA to send satellites and other equipment to high orbits. With ion-drive, low-thrust upper stages, SEPs may also be used for missions to comets, asteroids, and the ends of the solar system. (Courtesy Boeing Aerospace Company)

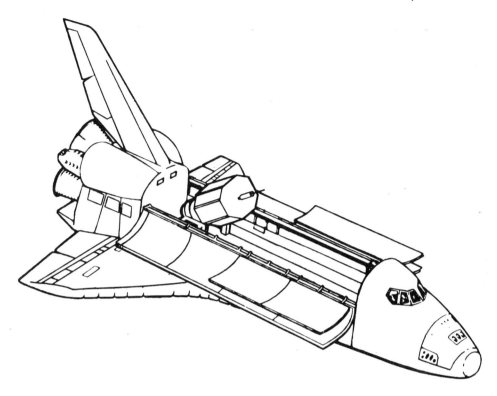

The solid spinning upper stage can place satellites and other material into orbits at locations and altitudes the shuttle does not reach. Here the tilting and spinning table is mounted in the cargo bay. (Courtesy NASA)

twin-stage; the NASA twin-stage, which is larger still; and a NASA three-stage spinner, for very high-energy interplanetary launches.

With the interim upper stage in the space transportation system, it is possible to send unmanned spacecraft and satellites up to geosynchronous orbit, which is called the Clarke orbit, after science writer Arthur C. Clarke who first suggested it. In a geosynchronous orbit, a satellite, space station, or spacecraft stays constantly over one place on the earth. Three communications satellites in the geosynchronous location, for example, can effectively cover almost all of the earth, if they are placed 120 degrees apart over the equator. The geosynchronous location is 23,500 miles from the earth.

IUS can also send unmanned spacecraft, or orbiting satellites, to the moon and to most of the other planets. Among the many hoped-for missions for space shuttle using an IUS interplanetary vehicle are a return to Mercury with a photographic reconnaissance craft and more probes; perhaps a lander to Venus; a return to Mars to place a *Viking*-type vehicle in several locations; unmanned orbiters for Saturn and its moon, Titan; and probes to visit Jupiter, of which *Galileo* is the first. Another much hoped-for mission is a sample-return mission to Mars. Soil and possibly biological specimens, always assuming they do exist because the results of the Mars *Vikings* 1976–1978 experiments were inconclusive, would be returned to earth's orbit for examination in space. This is one argument for a permanent space station at a later date: samples from other planets could be examined in space without the problems of quarantine which preceded the handling of the Apollo lunar samples. IUS may also be used to launch the Solar Polar mission to study the sun at close range, and perhaps the long-discussed Lunar Polar orbiter to chart the unmapped polar regions of the moon.

For solar system exploration, the Space Shuttle Transportation System also has the solar electric propulsion system, or SEPS. SEPS is an ion-drive propulsion unit designed by NASA for low-thrust, deep-space missions after shuttle has become operational. It can be used where a high fly-by velocity, such as combinations of IUS stages would provide, is not wanted or is not possible.

In the ion-drive SEPS, electrical energy is converted into the kinetic energy of the exhaust beam. Raw electrical power is obtained from a large solar array and converted by a power conditioner into the several different voltages required to operate an ion thruster. The thrust for the spacecraft stage is actually produced by the electrostatic expulsion of ions of mercury.

The solar electric stage is not a high-acceleration stage, but it has the virtue of being continuous. Once an IUS has put a spacecraft into escape velocity from the earth's orbit, the SEPS can thrust for as long as the fuel lasts, using solar power for the main propulsion. The eventual build-up of velocity using a solar electric stage could be extremely large, if one were designed to thrust continuously for five or ten years.

An empty SEPS weighs about 3000 pounds (1360 kilograms) and can fit easily into the space shuttle's cargo bay. Within its design lifetime of five years, a SEPS may have as much as twenty thousand hours of thrust time.

NASA-proposed missions that would use the space shuttle as the initial launch vehicle, an IUS as an interplanetary stage, and an SEPS as a final stage, include missions to the asteroid belt, where a slow fly-by would return much better scientific information. Other possible missions would chase comets, particularly Halley's comet on its return in 1986 to the vicinity of the solar system.

The shuttle/IUS/SEPS combination may also be the first set to push a spacecraft into the region of ultraplanetary space, the area between the solar system and a distance of about 0.1 light years. The time for information return, however, is still much too long for the distant missions, even with the best shuttle equipment and the most powerful interplanetary stages. Even Pluto, for example,

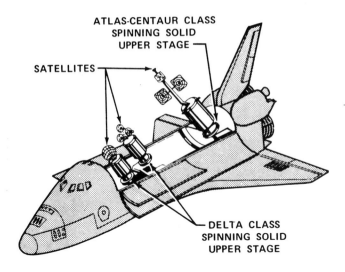

The solid spinning upper stage will place satellites of the Atlas-Centaur and Delta classes, based on weight and volume, in orbits near the orbiter. (Courtesy NASA)

might have an information return of more than forty years with the present equipment.

Parts for a large interplanetary mission would probably require two shuttle flights. One flight could bring up a heavy and complex NASA interplanetary IUS three-stage, which weighs in at 54,600 pounds (24,767 kilograms), nearly a full shuttle load. Another flight would bring up a heavy spacecraft attached to a SEPS vehicle. Together, a fully loaded sample-return and orbiter spacecraft with a SEPS would weigh perhaps 20,000 pounds (9072 kilograms). Shuttle would move it to the orbit of its previously delivered IUS together with other payloads designed for the same or a similar orbit.

When the interplanetary flight was completely assembled, it would be checked in earth orbit and then sent on to earth-escape velocity by its IUS. Then the SEPS would take over and it would begin its journey into the far reaches of the solar system.

A third Space Shuttle Transportation System propulsion device for use in space is the solid spinning upper stage or SSUS. It is also called a PAM, for payload assist module. This little jewel is manufactured by a private company and is intended to be used by space shuttle customers who will be placing satellites in altitudes or locations where the space shuttle does not go.

There are two sizes of SSUS/PAM: the SSUS-A/PAM-A is designed and sold to be used with those satellites which were formerly launched by the

Atlas-Centaur, earth-based, expendable, launch-vehicle system. The SSUS-D/ PAM-D is expected to be used for satellites which were formerly launched by Delta rockets from Cape Canaveral. Small communications satellites and small business satellites were previously Delta-launched, for example.

The SSUS is a solid propellant booster like the space shuttle's solid rocket boosters, and it will not be recovered for reuse. The smaller one, SSUS-D, is only 48 inches in diameter and 76 inches long (121.92 by 193.04 centimeters) and weighs about 4500 pounds (2038.5 kilograms). Both SSUS vehicles will be available for use with the space shuttle in the mid-1980s. The main use for the SSUS will be to place satellites and other craft in transfer orbit from low-earth to geosynchronous orbit.

Considering the planning and the relatively low budgets which have been allocated to space shuttle, its operation has very few limitations. One is the duration, which is about thirty days. At the thirty-day limit, the hydrogen-oxygen fuel cells of the space shuttle are used up and there is no more power. In addition, the cryotanks, which produce power for the extended shuttle missions, add 16,000 pounds (7257.6 kilograms) to the space shuttle weight, which could just as easily be payload.

Several studies have indicated the need for a power or utilities module for the shuttle. Such a module would give the shuttle solar-electric generated power for internal use, much like a portable utilities trailer used at some construction sites.

The module would be carried to low-earth orbit by the space shuttle and then deployed. It would supply plug-in utilities for extended missions. With the most-discussed design, a 25-kilowatt power module, which supplies 59 kilowatts under full sun, the shuttle would have enough power for a sixty-day mission, at which time the limitations of crew life support and food requirements would be reached.

In addition to supplying a shuttle with power for an extended mission, the power module could also be used to power free-flying payloads. Dual docking ports would allow the free-flying payloads to remain attached while the module is being attached to the space shuttle. In this way, the power module could support the operation of larger and more complex systems in orbit.

The power module would also substantially extend the mission time of the spacelab, which will be described in the following chapter. Current plans call for a baseline orbiter-spacelab mission of twelve days. With a power module, the spacelab could exist in space for up to sixty days or longer without cutting its payload volume or weight.

If a spacelab were to be modified into a free-flyer and it used a 25-kilowatt power module, it could remain in space indefinitely—if it were supplied regularly with food, water, and oxygen from the ground via space shuttle.

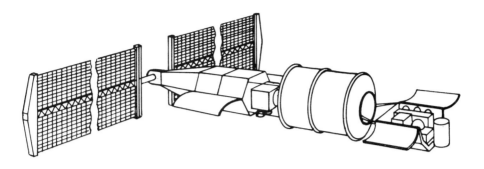

Spacelab with a 25-kilowatt power module can serve as a long-duration free-flyer. (Courtesy NASA)

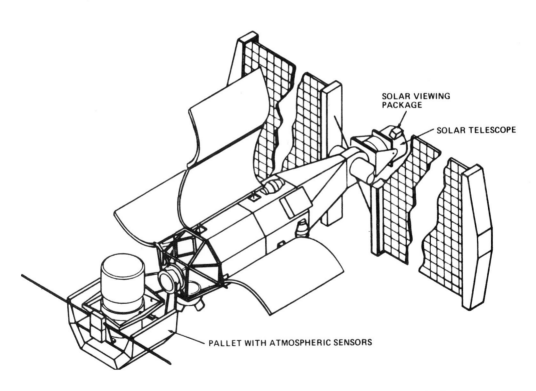

SOLAR VIEWING
PACKAGE

SOLAR TELESCOPE

PALLET WITH ATMOSPHERIC SENSORS

The 25-kilowatt power module can support free-flying payloads. (Courtesy NASA)

The power module is another piece of the space shuttle equipment which can pay for itself by commercial use. Many experiments require longer than thirty days in space. Observations of two solar cycles require fifty-six days, and biological experiments needing uninterrupted zero-g for successive generations of animal or plant life require variable periods. It has been estimated that a 25-kilowatt power module would pay for itself in the first year of operation. Power modules for space shuttle use may be ready for shuttle flights as early as 1984 or '85.

We are at a point in history where a proper attention to space . . . may be absolutely crucial in bringing this world together.

Margaret Mead

8

SPACELAB

The men and women move around in the small quarters of the spacelab. Here an experiment grows a crystal so pure it could never be found on earth. There a process is developed for a steel as light as wood and as strong as an I-beam. The pages of the electronic notebooks of the scientists have been filled with the new, untried, hoped for, unusual.

For the first time in the history of mankind, there are materials of the future to build structures of the future. The spacelab, at first going and coming with each shuttle flight, safe and locked in the cargo bay, has now become a free agent. It can stay in space for up to three months. One lately has been in space for six months, powered by an attached power module which gets energy from the sun. It is connected to a habitability module, home for the crews when they are not working on experiments and new processes for building the first construction base on which the permanent crews will soon make their living and their homes.

One of the big payloads for space shuttle will be spacelab. While the orbiter will provide the means to open the space frontier to additional and better satellites,

Spacelab in orbit. Two crew members are manning the experiments, while the orbiter crew is resting. The spacelab can remain in space inside the orbiter's cargo bay for missions of one week to one month. If it were separately powered in orbit and detached from the orbiter, it could begin extended missions of several months. (Courtesy NASA)

commercial uses of near-earth orbit by satellite, and cargo room for a hundred other uses, it is spacelab which will give mankind's scientific community easy, economical access to space and its future technology.

Spacelab is a basic, shirtsleeve working environment, with outside platforms called pallets on which experiments in space can be performed. For a given flight there are thirteen possible combinations of the lab and pallets, or of pallets alone.

Spacelab is a joint international effort between NASA and the European

Experiments such as the atmospheric, magnetosphere, and plasmas in space (AMPS) project will use the spacelab pallets. In this artist's view, the AMPS is contained on one pallet, while another has the accurate pointing system for the experimental project. Spacelab also has provisions for an airlock for experiments which need exposure to space, and it can be provided with a viewing window and viewport for some missions. (Courtesy NASA)

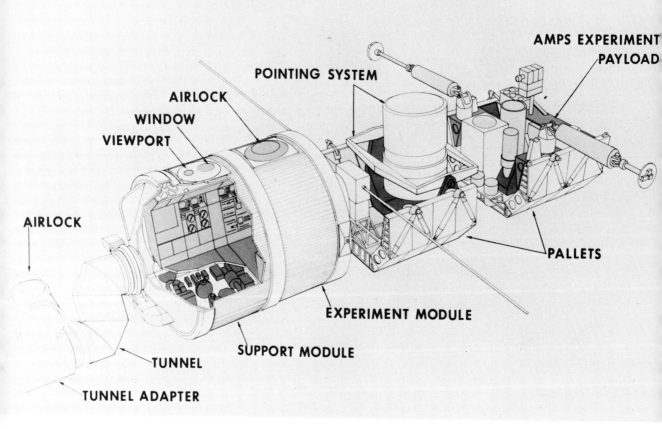

MODULAR SPACELAB

Space Agency. More than forty companies in ten countries are building the systems and subsystems of the first spacelab. Experiments expected to be aboard the spacelab will have completely international origins. For the first spacelab flight, seventy-seven experiments have been chosen. Of these, sixty-one are European, fifteen are from the United States, and one is Japanese.

The lab is designed to be launched in the orbiter's cargo bay and remain there during a mission. The mission duration for spacelab is the same as for the orbiter: seven to thirty days. In most of the thirteen flight configurations, the lab will be connected to the orbiter's airlock by a tunnel, leaving a free passage between the two. As with other Space Transportation System components, the lab is reuseable and modular.

The basic spacelab is a pressurized cylinder 22.5 feet (6.9 meters) long and 13.7 feet (4.2 meters) in diameter. When the mission requires it, two modules can be easily joined together to make a long spacelab. Up to five of the pallet sections can be carried on a mission. The pallets serve as mounts for instruments such as telescopes, sensors, antennas, and experiments which will be exposed directly to the vacuum of space. The equipment on the space pallets can be operated remotely from inside the pressurized module of spacelab, or from the payload-mission specialist's station on the aft flight deck of the orbiter.

The concept of a spacelab goes back to the old ideas of a permanent, orbiting space station, where the uses of space as a new environment for science and engineering were first explored. A permanent space station, of course, could also monitor the earth and so can the spacelab. Inside the pressurized module of spacelab, in a standard sea-level atmosphere, the crew can study the earth's weather, pollution, water, and other resources. They can conduct experiments with new and old vaccines, semiconductors, glasses, and alloys. Communications and navigation experiments can be performed, as can medical and biological research projects into life processes. In the field of physics, the spacelab's compartment can be used to study atmospheric and cloud physics, and cosmic phenomena.

Outside, on the space pallets, detectors and sensors for stellar, solar X-ray, infrared, and ultraviolet studies can be mounted. Experiments which require large antennas for fine-pointing control can also be mounted on the pallets.

Leaving aside the older designs for a permanent station, which contained many of the same ideas, the lab was originally studied as a design concept in 1972 and was known as Sortie Can. When the idea was finalized, spacelab became one of the major steps in sharing the costs of the utilization of space. The entire cost of the spacelab is borne by nine European Space Agency countries: Belgium, Denmark, France, Germany, Italy, the Netherlands, Spain, Switzerland, and the United Kingdom. Austria is also contributing to the cost of building the spacelab, although it is not a member of the ESA.

NASA establishes the program requirements, builds and maintains support

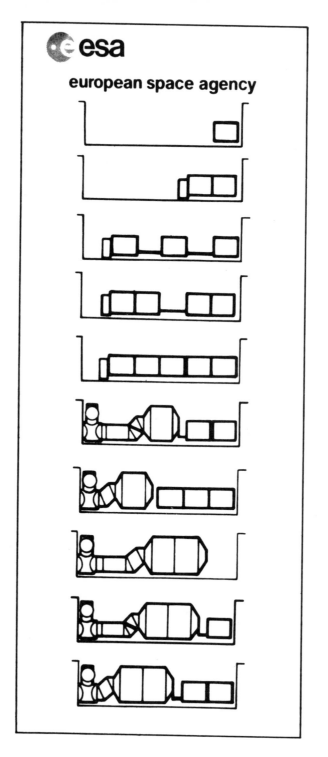

Spacelab will use a variety of standard configurations of pallets and habitable modules on different missions. (Courtesy European Space Agency)

The spacelab mockup in Building 36 at the Johnson Space Center. In 1977, week-long missions were flown in the simulator. (Courtesy Johnson Space Center and NASA)

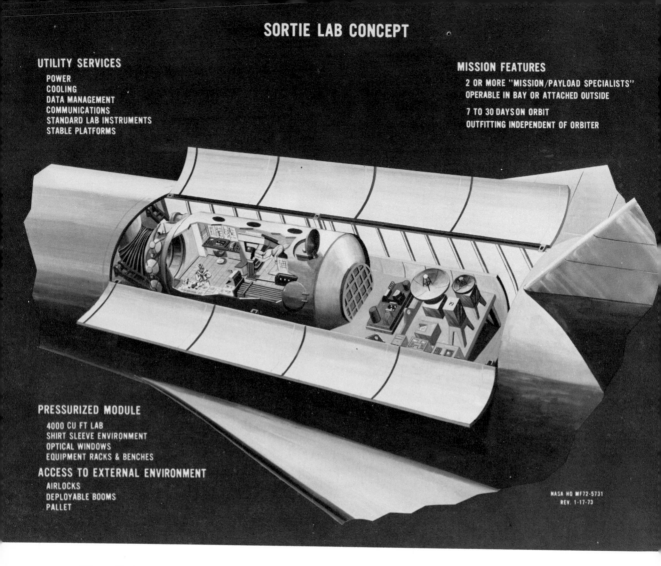

SORTIE LAB CONCEPT

UTILITY SERVICES
POWER
COOLING
DATA MANAGEMENT
COMMUNICATIONS
STANDARD LAB INSTRUMENTS
STABLE PLATFORMS

MISSION FEATURES
2 OR MORE "MISSION/PAYLOAD SPECIALISTS"
OPERABLE IN BAY OR ATTACHED OUTSIDE

7 TO 30 DAYS ON ORBIT
OUTFITTING INDEPENDENT OF ORBITER

PRESSURIZED MODULE
4000 CU FT LAB
SHIRT SLEEVE ENVIRONMENT
OPTICAL WINDOWS
EQUIPMENT RACKS & BENCHES

ACCESS TO EXTERNAL ENVIRONMENT
AIRLOCKS
DEPLOYABLE BOOMS
PALLET

NASA HQ MF72-5731
REV. 1-17-73

The preliminary concept for the spacelab was called Sortie Can in 1972. (Courtesy NASA)

equipment, and trains experiment operators, in addition to operating, maintaining, refurbishing, and overhauling the spacelab. The best part of the deal, of course, is that the United States will be able to *use* the spacelab and to buy additional spacelabs from the ESA for completely United States-oriented programs.

The eight basic forms of a spacelab mission use one or two pressurized modules, called long modules, and two or more pallets for scientific experiments. If only the long module is flown, it provides the most room, 784 cubic feet (22.2 cubic meters). The long module with one scientific pallet provides maximum pressurized room, with an outside experimental platform of 184 square feet (17 square meters). Other variations are long module plus two pallets and short module plus two or three pallets. The scientific pallets may also be flown by themselves, without the spacelab pressurized module. Variations include one to five pallets in a train in the cargo bay of the space shuttle.

If only the scientific pallets are used, a cylindrical igloo on the end of the forward pallet provides services normally coming from the spacelab, such as electricity and temperature control.

The spacelab interior is extremely compact, as might be expected, with racks for instruments and experiments of different flights. The lab can support three payload specialists working a twelve-hour shift, and four can be accommodated

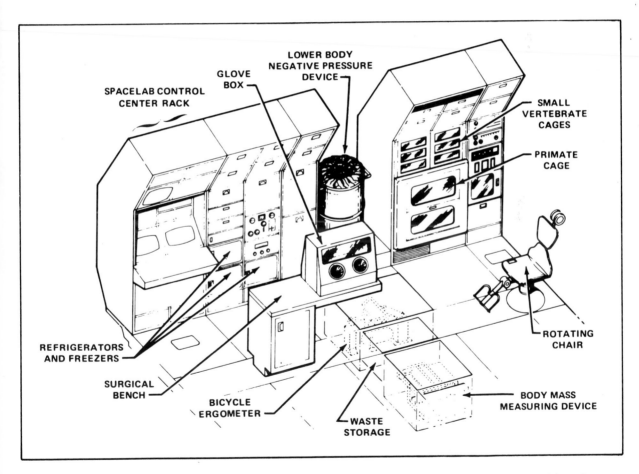

Typical layout for a spacelab dedicated to life sciences, with possible locations of center aisle and starboard racks. (Courtesy NASA)

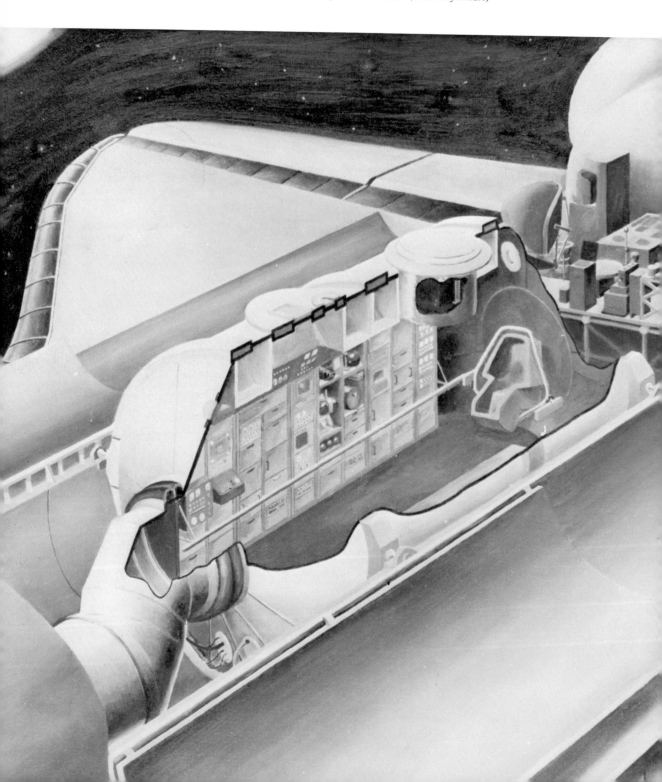

for a one-hour period of shift overlapping. The lab temperature is maintained between 64 and 81 degrees (17.8 to 27.2 Celsius). Throughout the spacelab, there are handholds, foot restraints, and mobility aids of all sorts, so the crew members can do all of their tasks as safely and efficiently as possible. As might be expected, there is a work bench inside the spacelab, complete with writing instruments and paper. And on each flight there are floor, wall, and ceiling containers for tools, personal belongings, and pieces of standard, spacelab-maintenance equipment.

Through the transfer tunnel, the spacelab crew can pass freely between the orbiter and the lab. The tunnel is large enough so that a crew member wearing an extravehicular activity suit can leave the orbiter and move into the spacelab or to the airlock in the docking module for outside activities.

For some flights, the spacelab will be equipped with a top airlock. The top airlock is 3.28 feet (1 meter) in diameter and length. The top airlock is specifically for experiments aboard the spacelab which have to be exposed to space environments and which are smaller than those on the scientific pallets. Once the outer airlock door is opened, the airlock experiment platform can be extended up to about three feet (1 meter).

The spacelab's optional optical window and viewport were adapted from the *Skylab*. The main optical window is 16 by 21 inches (40 by 53 centimeters). The viewports, if installed, are 11.81 inches (30 centimeters) in diameter. If the spacelab mission does not call for airlock experiments or windows to view experiments, the locations are blocked with a cover plate.

One proposed spacelab project is the astrophysics payload. This involves a set of instruments to investigate the life cycle of the sun and stars, evolution of the solar system, and the origin and future of matter. Using an all-pallet payload system, the seven-day mission has also been called the high-energy astrophysics observatory.

Another astronomical spacelab is the solar physics payload. This project would view the interactions of the sun and the earth, especially investigating the solar wind and high-energy particles. The SPP is also an eleven-pallet possibility, controlled either from the payload station of the orbiter or from the ground.

The astrophysics-pallet spacelab flights would orbit at an altitude of 216 nautical miles (400 kilometers), while the solar-physics scientific pallets would be in a higher orbit at 254 nautical miles (470 kilometers). The astrophysics flight would be a seven-day mission; the solar physics spacelab trip would be longer—about fifteen days.

A manned spacelab flight to study physics has been named the atmosphere, magnetosphere, and plasmas in space project—AMPS. This spacelab project would investigate the earth's electric and magnetic field system, particle and wave interactions, and the physical properties which are associated with the motion of

bodies in very thin plasmas. It would also study the chemistry and dynamics of the upper atmosphere of the earth.

Other projected spacelab projects include an advanced technology laboratory, in which researchers into future space technology would have complete vacuum, zero gravity, and the ability to extend and recover experimental equipment. The ATL could, for example, be composed of a long module and two scientific pallets, and would orbit for a seven-day mission.

An additional spacelab project is the earth-viewing application laboratory. This spacelab project, among other things, would perform a world crop survey, assist urban planning, survey minerals, inventory water, study the weather and climate, and investigate the earth's oceans.

A life-sciences spacelab has been discussed for future missions. In seven-day, and later in thirty-day, flights, a life sciences-biology spacelab would study the effect of the space environment on plant and animal tissue, cells, bacteria, and the whole range of possibilities which space environments have presented to biology and medicine. Very high priority has been given to designing experiments for the investigation of heart and respiratory diseases, and to medical problems which are related to space travel and living for extended periods of time in space.

Among the devices planned for investigations into the medical aspects of space travel is the spacesled. Spacesled will be used to research space sickness, which seems to be approximately the same as the more familiar motion sickness, consisting of spells of dizziness, profuse perspiration, and general nausea. A payload specialist aboard the spacelab will be strapped into the sled and subjected to special acceleration simulations. Breathing, pulse rate, eye movements, and other physical activities will be measured to determine the causes of space sickness.

Aboard the spacelab, as a piece of standard equipment, will be a materials sciences laboratory to be used for delving into new techniques of manufacturing and processing in space. The materials laboratory will have a device for metallurgical experiments on materials with melting temperatures up to 1500 degrees Celsius. A high-temperature thermostat has been developed for research into molten metals in space, and there will also be facilities for growing protein single crystals under zero-g conditions.

Space has incredible possibilities for new manufacturing processes. The main reason is the complete lack of gravity. Metals, for example, can be solidified without the effects of convection, which is gravity-derived. In zero gravity crystals can be grown which have improved electronic characteristics and are distributed

Another experiment pallet in the spacelab and the orbiter in low-earth orbit. Spacelab will be constructed by the European Space Agency, although Americans will use the facility. There are provisions for the United States to purchase spacelabs for its own use. (Courtesy NASA)

more homogeneously than is possible on earth. Space also offers a high vacuum where evaporative purification research can be performed which on earth would be extremely expensive.

On earth, dense materials drop to the bottom of containers, less dense materials float to the top. Gravity pulls molecules apart, leaving holes for impurities and causing defect lines which reduce strength or change electrical properties. None of this exists in space. High-strength alloys are easily possible in space, leading to turbine blades which are much more efficient and longer lived than those manufactured today. Permanent magnets may be made in space to replace the energy-consuming electromagnets now used in electrical equipment. Amazing new ceramics and glasses, which are solidified without containers in outer space, will be a product of experiments aboard the space transportation system of the future. In a dozen fields, a world with an entirely new environment is opening up above the earth's atmosphere, in the absence of gravity. No one can predict what will eventually happen when gravity is removed from mathematical equations for properties of materials.

Materials-processing experiments have already been conducted to a limited extent in space by the crews of America's *Skylab*. The Russians have been several steps further with space factory research in their *Salyut* space station. Russian cosmonauts have performed over fifty technological experiments with their Splav and Kristall space furnaces.

The first of the spacelabs, Spacelab 1, was originally scheduled for launch in July 1980. Subsequent delays have pushed the launch date into 1981. The flight will be launched at Kennedy Space Center to an earth-orbit height of 155 miles (250 kilometers). The first mission is scheduled for seven days, with a variety of experiments, including stratospheric and upper-atmospheric physics, materials processing, space-plasma physics, life sciences, astronomy, solar physics, earth observations, and space technology. The Spacelab 1 launch will have a crew of five payload specialists, of whom two will be Americans. Two of the payload specialists will fly aboard the spacelab and operate the science instruments.

Spacelab 2 is scheduled for launch in the spring of 1981 with a payload of mostly astrophysics experiments. It will be an all-pallet mission, with no pressurized module. Among the experiments to be flown are a small infrared telescope, a cosmic-ray experiment, two life-sciences experiments, and two experiments in space technology.

Spacelab 3, tentatively scheduled for a launch toward the end of summer in 1981, will also have a varied group of experiments. The early shuttle-spacelab flights will, in addition to pioneering in a dozen fields, be testing the shuttle and

The AMPS payload for shuttle/spacelab is shown here in earth orbit. On the planet's surface, the larger cities light up the darkened landscape. (Courtesy Marshall Space Flight Center)

the spacelab equipment to find problems with the system. If the schedule holds, later in 1981 the shuttle will be used to fly a series of spacelab pallet missions. One is scheduled in space processing, one in life sciences, and another in physics and astronomy.

What the future may hold for spacelab is, to some extent, dependent on budgets and politics. The spacelab, as other Space Transportation System vehicles, is a reuseable, modular, highly adaptable unit which, in turn, means lower costs for anyone wanting to experiment in space.

One of the first of the long-range plans made for the spacelab was an extended mission configuration which would allow missions up to several months for the crew of the orbiter-spacelab. Looking into the future, all of the facilities aboard the spacelab can be duplicated aboard a permanent space station. It is through the possibility of initial research and investigation aboard the spacelab that the technology will be created to build America's first permanent manned station in earth orbit.

The universe is not only stranger than we suppose, it is stranger than we can suppose.
John B. S. Haldane, 1928.

9

ASTRONOMY IN SPACE

The galaxy has witheld too many secrets. It was only fifty years ago that man realized that all of the stars, globular clusters, open clusters, and nebulae did not belong to the near reaches of the Milky Way galaxy. It is hardly four hundred years since man knew of no moons around other planets, no ice caps on Mars, no hated hell of temperature and pressure on Venus.

In high earth orbit, the Space Telescope probes into the space far beyond distances easily imagined only by friendly computers. The speeds of stars are measured, their temperatures are checked, the ways in which they approach and recede from the vicinity of the solar system are noted. Above the atmosphere, which absorbs so much of the light and energy from the stars, the Space Telescope and a dozen special satellites keep watch on the universe. They constantly remake the theories and the viewpoint of the scientists who have designed them.

As mankind begins the long migration from the earth's surface toward the stars the telescopes see, the instruments interpret information that changes the picture. The space shuttle brings more and more of them to the orbits and sends more probes to and beyond the edge of the solar system. Already the probes have landed on Mercury, encountered Uranus and Neptune, flown by comets, and landed near the surface of selected asteroids. Man is going outward from his base aboard the space shuttle. A hundred men and women have been into space, a thousand more will follow.

The dream of a telescope in space, far above the turbulent and obscuring atmosphere, is as old as the first dreams of extensive use of near-earth orbit by space station. Hermann Oberth first proposed a telescope in space in 1923. Other authors also proposed space telescopes, including von Braun in the *Collier's* magazine series, "Man Will Conquer Space Soon," and in his 1952 book *Across The Space Frontier*.

Astronomers had been sending up small telescopes and other devices in high-altitude balloons and upper-atmospheric sounding rockets for years, but it was not until the mid-sixties that satellite astronomy became possible. The first orbital astronomical observatory (OAO-1) was launched on April 8, 1966. It carried a 16-inch (40.94-centimeter) telescope and four 8-inch (20-centimeter) telescopes. Unfortunately, it stopped operating after only two days because of a power failure.

OAO-2 was launched more than two years later on December 7, 1968. This satellite-observatory carried a group of four 12½-inch (30-centimeter) telescopes for ultraviolet observations of very hot stars, known as main-sequence stars. It also carried a 16-inch (40.94-centimeter) reflecting telescope for studies of nebulae and four 8-inch (20-centimeter) telescopes for other studies. It was successfully launched into a near-earth orbit.

OAO-B was a failure. The spacecraft fairing failed to jettison and the satellite never reached orbit. It carried an ambitious project: a 36-inch (91-centimeter) telescope.

The *Copernicus* satellite, also known as OAO-3, was successfully launched on August 21, 1972. It achieved orbit carrying a 32-inch (81-centimeter) telescope plus three X-ray detectors. All of this space activity with small telescopes over a period of ten years only served to demonstrate the need for a truly large telescope in orbit.

Discussions about a large space telescope began in 1962 and 1965. The National Academy of Science's Space Science Board organized studies, discussions, and committees in those years and continued the work until 1969. The result of the studies was a wide acceptance of the idea of a large space telescope in

The large Space Telescope will have images more than ten times finer than those of earth-based telescopes. It is scheduled for launch aboard space shuttle in the 1980s and should have a minimum orbital lifetime of ten years. It can be maintained and repaired in space or brought to earth for repairs and remodeling. (Courtesy Marshall Space Flight Center and NASA)

near-earth orbit. Detailed engineering studies were conducted by NASA, ending in the designs approved by NASA in 1977 and approved by Congress in 1978.

The Space Telescope will be launched aboard the space shuttle in the last few months of 1983. It has a 96-inch (2.4-meter) mirror and will orbit the earth at a planned altitude of 310 miles (500 kilometers) at an inclination to the earth's equator of 28.8 degrees.

Not only will the shuttle launch the Space Telescope from its cargo bay, but shuttle will be used as a base from which astronauts can make repairs and replace the instrument packages aboard Space Telescope for new experiments. If it becomes necessary, space shuttle can pull the Space Telescope into its cargo bay and return the instrument to earth for extensive repairs, maintenance, or general overhaul, which shuttle has been scheduled for about 1988.

The Space Telescope has a length of 43 feet (13.1 meters) and a diameter of 14 feet (4.26 meters). It fits easily into the space shuttle's cargo bay and its weight accounts for about one-third of a payload. Electrical power to operate the telescope will come from batteries which are charged by large solar panels on the sun side of the Space Telescope's orbit. The images received will be transmitted to the earth by telemetry.

The guidance system of the Space Telescope is so accurate that it can hold on target within 0.007 arc/sec. This is an angle only slightly larger than that made by a dime when viewed at a distance of 400 miles.

The open end of the Space Telescope will be similar to most telescopes on earth. The light from the stars, or other objects being studied, enters the open end of the telescope and strikes a reflecting mirror at the rear of the instrument. The reflecting mirror then projects the image back to a smaller mirror in the front. The beam of light is reflected back down the instrument to the scientific instruments in the rear. This is a modification of the traditional Cassegrain telescope form called a Ritchey-Chrétien.

The Space Telescope will be much more powerful than the largest earth-based telescope in the United States, the 200-inch (5-meter) one on Mount Palomar in California. Although the Space Telescope is only one-half the size of the Palomar giant, it is expected to sweep a volume of space more than *three hundred and fifty times* that of the larger instrument.

Hampered as it is by the earth's atmosphere, even though it exists in a location of some of the steadiest air in the world, the 200-inch Mount Palomar telescope can see out to an estimated two billion light years. A light year is six trillion, trillion miles. Telescopes on earth are restricted by two major factors: the long column of turbulent air through which they look, and their physical size,

Opposite: An interim upper stage, attached to the payload it will propel to high earth orbit or to another planet, drifts away from orbiter while another prepares to be released from the manipulator arm. (Courtesy Boeing)

Left: This free-flying power station can support an orbiter for extended flights of up to 120 days. (Courtesy NASA)

Each shuttle flight will carry enough life support for two six-hour extravehicular activities. Below: high above launch point in Florida, a long duration exposure facility receives an experiment tray. (Courtesy NASA) Opposite: crew members repair a satellite. (Courtesy Rockwell International Space Division)

Opposite bottom: Shuttle bay doors have been opened to expose Spacelab's pressurized work compartment and pallets containing scientific instruments. (Courtesy NASA)

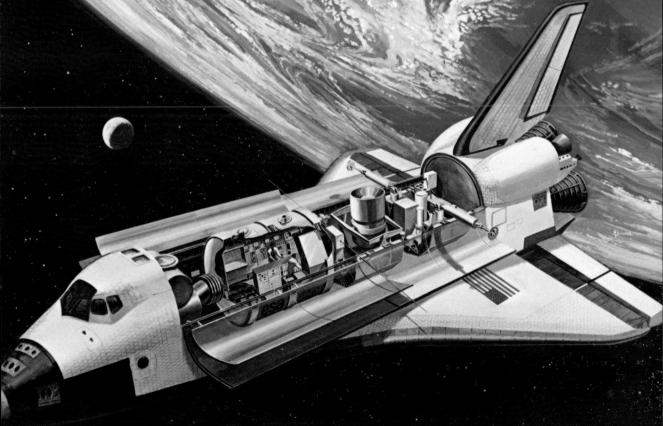

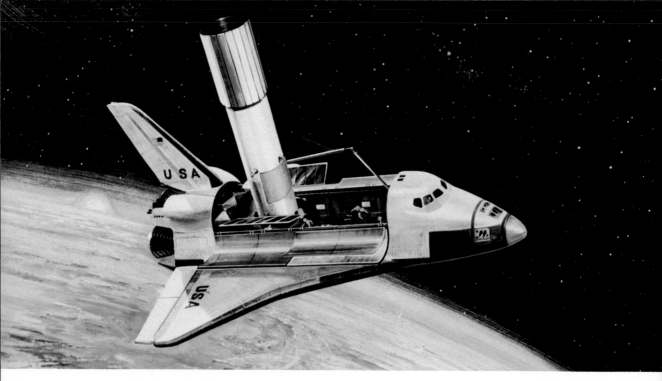

Approved design for the large space telescope. (Courtesy NASA)

A film cassette is inserted into space telescope. For the operation, work lights are provided on the manned maneuvering unit, and the crew uses foot rests built into the space telescope structure. (Courtesy NASA)

In this artist's concept, the shuttle crew is preparing to release the large Space Telescope from the orbiter payload bay. (Courtesy NASA)

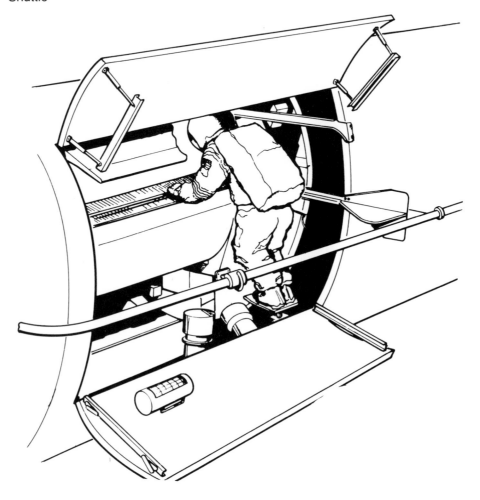

An astronaut services the Space Telescope in this sketch. The shuttle orbiter can be used to carry astronauts and technicians from earth to the orbit of the space telescope. (Courtesy Martin Marietta Aerospace Denver Division)

which is limited by gravity. A very large telescope mirror will sag due to gravity. Its perfect optical shape will distort and it will be useless. In addition to being turbulent the atmosphere filters out and shields the earth from certain wavelengths of electromagnetic radiation, of which visible light is only a part. Space telescopes can detect wavelengths in the infrared, ultraviolet, and X-ray regions, and everything in between. A space telescope, or any object in space, is exposed to the entire range of the spectrum. If it can be designed at all to detect a certain radiation, it can detect that radiation in space.

Space Telescope can photograph objects a much longer time than earth-based telescopes, resulting in images up to ten times finer showing much fainter stars. Just in the immediate solar system, the Space Telescope will probably show undiscovered moons—tiny ones never before seen because of the earth's atmosphere—around Jupiter, Saturn, and perhaps, Uranus and Neptune. It can show more about the strange and newly discovered moon orbiting the distant Pluto.

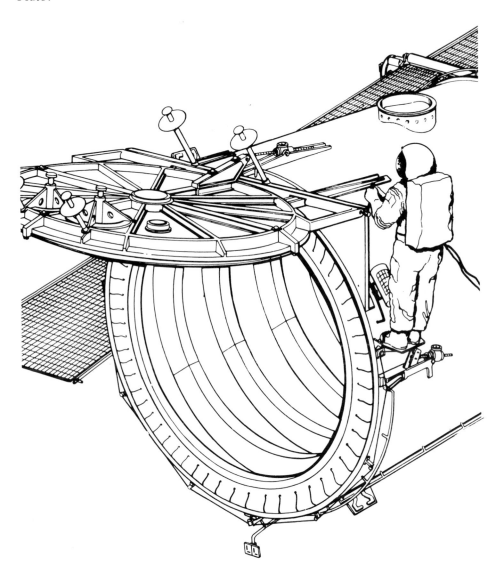

A shuttle crew member in an EVA suit is shown here working on the space telescope. (Courtesy Martin Marietta Aerospace Denver Division)

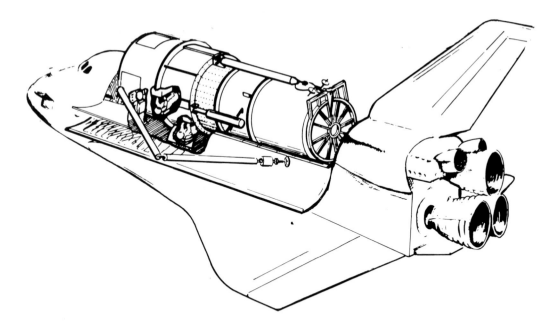

The space telescope tucked away in the shuttle orbiter cargo bay. The space telescope is one of the most important payloads for scientists. It is expected to reveal extensive new information on the origin of the solar system and the creation of the universe. (Courtesy Martin Marietta Aerospace Denver Division)

But the Space Telescope will focus on more distant targets. It will be used to look at quasars, galaxies, gaseous nebulae, and Cepheid variable stars. Objects such as these, which are fifty times fainter than what can be seen now, will be within the range of the Space Telescope. It will be used almost 5,000 hours each year for observation, while earth-based telescopes are limited to 2,000 hours per year at best. It can even make some observations, although not of the faintest subjects, while in the sunlight.

One of the most exciting projects for the Space Telescope is a search for planets orbiting nearby stars. Although ground-based photographic evidence and mathematical inferences indicate that there are several nearby stars with Jupiter-size planets, the Space Telescope can confirm this type of information. It may possibly detect planets as small as Earth and Mars going around stars in the sun's immediate neighborhood.

Space Telescope, as a 1983 payload, will consist of the main instrument with solar panels and a modular equipment package, plus two cameras, two spectrometers, and a photometer.

The Space Telescope will not be the only astronomical instrument which the shuttle will take into space. A planned 1984 mission will place a gamma-ray satellite in orbit. Gamma rays are the most energy-packed form of radiation so far discovered. The satellite will be equipped to detect gamma-ray bursts, nuclear gamma rays and very high-energy gamma rays from pulsars.

The kind of information which a gamma-ray satellite can collect should lead to a deeper understanding of the nature of supernovas, pulsars, quasars, and radio galaxies. A supernova is a large star at its life's end. Its final collapse is a cataclysmic event that generates a violent explosion. The last supernova in our own Milky Way galaxy was seen in 1604. Pulsars were discovered in 1967, emitting radio signals with extremely precise pulsation. The evidence seems to

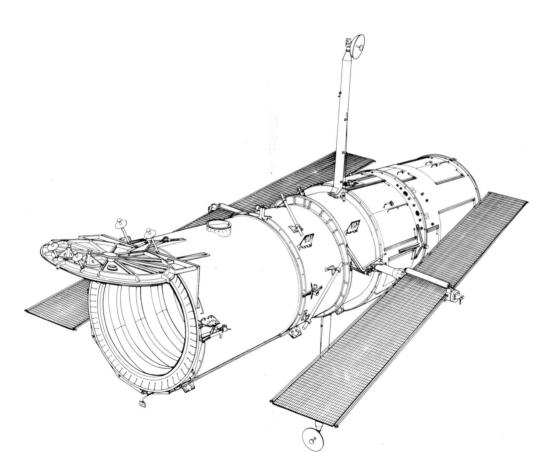

One of the space telescope designs which was studied by NASA is shown in this sketch. (Courtesy Martin Marietta Aerospace Denver Division)

suggest that pulsars are fast-spinning neutron stars, which are compact bodies of densely packed atomic particles having no electric charge. Neutron stars are thought to form when a very large star burns up its fuel and collapses. Quasars, for quasi-stellar-radio-source, are objects which look like stars, but which emit more radio energy than the most powerful galaxies known. And they seem to be incredibly distant. If they are as distant as some astronomers seem to think and the energy is as great as it appears to be from the earth's location, then the amount of energy emitted by a quasar in *one second* could supply all of the earth's electrical energy needs for a *billion years*.

The gamma-ray satellite may also shed some light on gamma-ray bursts, which are erratic and appear every month or so, flashing across the solar system. Like X rays and ordinary light, gamma rays are a form of electromagnetic radiation, or photons. Unlike ordinary light, gamma rays have extremely high energy and a very short wavelength. A first class gamma ray packs several million times as much energy as the same unit of ordinary visible light.

The study of gamma rays in space is quite recent. They were recognized just after World War II, but it was not until 1967 that experimentation with a Vela satellite confirmed their existence. Gamma rays are absorbed by the earth's atmosphere and cannot be studied except by satellites and rockets. Following the formal discovery of gamma rays in 1967, they were detected from instruments aboard the later series of Orbiting Solar Observatory satellites, which OSO launched between 1962 and 1975.

In 1972 a gamma-ray detector satellite called Small Astronomy Satellite or SAS-2 was launched. Another, known as COS-B, was launched in 1975. Further gamma ray observations were made by the High Energy Astronomical Observatory launched on August 12, 1977 which also measured X-ray sources in the sky. More gamma-ray studies will come from HEAO-2, launched in November 1978, and HEAO-3 to be launched in 1979, as will more studies in X-ray astronomy.

All of these previous studies paved the way and led to the design of the gamma-ray satellite, GRO, for Gamma Ray Observatory.

There is little question that the operational years of space shuttle will see possibly dozens of satellites whose main purpose is the study of astronomy. With the Space Telescope and the versatility of the Space Transportation System, the world of astronomy may be as greatly expanded as it was when the first telescope was turned on the skies almost four hundred years ago.

*The **Galileo** orbiter probe will be launched aboard the space shuttle in 1982–83. It will orbit the planet for at least twenty months after a flight of one thousand days. Atmospheric probes will be dropped into Jupiter's atmosphere from the **Galileo** orbiter. (Courtesy NASA)*

In addition to placing the Space Telescope in orbit, and transporting and servicing a large number of astronomical observatory satellites, the space shuttle will also contribute to the field of planetary astronomy by launching interplanetary spacecraft, for which it will use the interim upper stage and the solar electric propulsion system.

Project Galileo, or Galileo Probe, will be an orbiter plus atmospheric-entry vehicle to be launched aboard space shuttle. The original Galileo Probe concept called for a launch in January 1982, with arrival near Jupiter in December 1984. Delays in the project may put the target date for Jupiter arrival nearer 1986. Because of tight funding, the Galileo Probe is a single-launch project with no spacecraft to back it up in case anything goes wrong. NASA has gambled this way before when the funding for a mission was tight, and the possible returns too great and interesting to ignore. *Mariner* to Mercury was a one-flight spacecraft which paid off the gamble and produced some of the best views of a planet ever taken by an unmanned probe. It returned to the vicinity of Mercury three times before it finally failed because of a shortage of attitude-control gas.

Galileo will arrive near Jupiter and release its atmospheric probe fifty-six days out from planetary encounter. The probe will enter the Jovian atmosphere and will broadcast until it is crushed by the pressure or burned by the high temperatures. Hopes are that the atmospheric probe will last at least fifteen to thirty minutes after atmospheric entry. The second part of the Galileo mission, the orbiter, will continue onward toward Jupiter after releasing the probe and will assume orbit around the giant planet. For the next twenty months, *Galileo* orbiter will survey Jupiter from a series of elliptical orbits covering the entire surface. It will also have eleven close encounters with the Jovian moons, Ganymede and Callisto.

Galileo is only the first of what should be a long series of interplanetary flights using the space-shuttle technology. During the next twenty years of the operational phase of the space shuttle, planetary astronomy will grow as it has perhaps never grown before. Fly-bys will go over the solar system; landers will visit the nearby planets; orbiters will take up positions, as *Mariner-9* did over Mars, and map surface features.

With the technology of the Space Transportation System, the planets will be studied by unmanned spacecraft as they could never have been studied from earth-based launch vehicles, partly because of cost, partly because shuttle will

*In the future, space shuttle may launch a complex mission to the planet Mars. The mission would include orbiters similar to the 1975–76 **Viking** orbiters which would control automated rovers on the surface of the planet. With the space shuttle, unmanned planetary exploration is wide open. (Courtesy NASA)*

allow more complex and ultimately more powerful flight combinations to be assembled in earth orbit.

The results of all this industrious tinkering with the secrets of the solar system should be a firm scientific knowledge of the space and planets around us, and a good base for planning the manned expeditions to the nearby planets, which are sure to come in the future with man's expansion into the dark above our planet.

I am probably the only one here who has done duty in a space station. Gentlemen, the costume we are wearing is customary in a station. A man fully dressed would stand out like an overcoat on the beach.

Robert A. Heinlein, The Puppet Masters

10

MODULES, SPACE PLATFORMS, AND SPACE STATIONS

It is finished now. The shuttle has brought the new materials and ways. The men and women have donned suits of foil and cloth to translate themselves in minute orbits around their own tools and machines that are centered around an imaginary point above the earth. The space station rides completed, waiting for the warmth of power, the rush of air, and the shuffling of uniforms.

On the earthside of the space station is the breakfast room, the Earthlight Room, named from some forgotten science fiction story. Across the station is a great Moon Room, built to hold twenty people at dinner. The names are only convenience; as the station rotates to provide artificial gravity, either dining room could be above the earth or looking at the moon at one time.

Attached to the docking end of the space station is a space shuttle orbiter. New and fresh from the shop, this is an improvement on the old shuttle, but looks much the same. It has brought from earth neither tools nor luxuries. It bears boxes, crates, and cartons which can be pushed into the waiting locks of the space station. An easy shove in the weightlessness brings them up against a catch net which will stop the flow of materials into the new home for astronauts.

Below and a few hours away, the first crew of the space station is waiting at the Cape for the word that the supplies have arrived and the station is ready. They will be the first men and women to take up permanent residence in space, staying above the earth until they are replaced by a relief crew.

Someday there will be a larger station, then still a larger one. There will be construction bases and terminals in space. In the nearing future, a spidery framework will be erected in orbit around the moon.

Space stations have flitted in and out of both popular and technical literature for about a hundred years, as was mentioned in chapter 2. The two most famous space-station concepts are von Braun's station described in *Across the Space Frontier,* and the giant wheel-station called Space Station Five in the film *2001—A Space Odyssey.* The von Braun proposal, in 1952, envisioned a space station 250 feet in diameter with a crew of about eighty. A *2001* station was a 1000-foot (305-meter) giant with a correspondingly large crew.

Of the two serious space-station proposals made during the age of space in the United States, both went down before severe budget cuts and lack of Congressional interest. About 1965, the United States Air Force developed the two-man orbiting laboratory concept; in 1969, the 12-man.

Testing concepts and information on permanent manned orbital vehicles was carried out by the Russians with their *Salyut-1* in 1971 and aboard America's *Skylab,* placed in orbit May 14, 1973. *Skylab* crews produced a record habitation of 89 days, while the Russians have recently set a new record of 139 days aboard *Salyut-6.* The Russian crews were supplied by unmanned cargo rockets.

There will be a race of sorts to complete the first permanently manned space station. The Russians have already indicated plans for a large station constructed of modules which would be assembled in space by the Progress space tugs. A 1980s Russian plan would involve a station with a crew of twelve to twenty-four. The station would be supplied by Progress unmanned cargo ships. Crews would arrive and depart via Kosmolyot, a reuseable spaceship and space plane similar to the American shuttle.

With the Space Transportation System, the United States has an excellent base from which to build large space structures such as space stations. And if the political winds shift sufficiently for the NASA budget to become flexible enough to do so, stations will be built, eventually.

There are several possibilities for an American space station in the near future without extensive funding. The first concept is an extension of the orbiter, or orbiter-spacelab, for longer missions. This can be accomplished with a power module, like the one described in chapter 7. With a 25-kilowatt power module attached either to the orbiter or an orbiter-spacelab combination, the space shuttle could remain in space for up to sixty days.

Since the orbiters of the Space Shuttle Transportation System are not among the more expendable items, it would be simpler to modify a spacelab into a free-flying unit, complete with a power module. This would, in effect, make a permanent space station, which could be supplied by orbiters. If a decision on a permanent station is made by 1982, then a spacelab with power module could be fully operational by 1986.

All of the plans mentioned are for a relatively old concept, a manned orbiting laboratory with either scientific or military priorities. What have become potentially more important in the last few years are the ideas of a manned construction base, which would facilitate experiments with space manufacturing, and the building of large space construction for the commercial utilization of space.

Studies completed in 1977 by NASA's Advanced Programs Division of the Office of Space Transportation Systems were aimed at what was described as a "long-duration manned orbital system," later called a "space construction base" since that was a more accurate description than the older name of space station.

One concept for the space construction base would be operable by 1984. At the beginning it would allow continuous operation in low orbit for a crew of four to eight, with room for expansion in the future. It would be assembled from modules, delivered to low-earth orbit by the space shuttle. Part of the concept includes a 25-kilowatt power module, with later designs including a 250-kilowatt photovoltaic power module, using solar power. Nuclear power is also a possibility for providing the electrical requirements of construction in space.

With a 250-kilowatt power module using solar power, the expanding structure could be called a Shuttle-Tended Space Construction Base. Such a construction base platform would probably have enough power to begin space construction through the middle years of the 1980s. If it were necessary to have the facility manned for long periods of time, or permanently, it could be expanded by a space-habitability module, shipped up in the cargo bay of the space shuttle.

Even while a modified spacelab operated as a space station in low-earth orbit, the space construction platform, or base, could be under construction for use in the late 1980s with a habitability module for the 1990s. Prior to the building of the module, the base would be tended by shuttle missions.

In the near future, then, there are several combinations of potential space stations: spacelab for thirty-day missions, spacelab with power module for sixty-day missions, and finally a space construction base, tended at first by shuttle, then perhaps by an attached modified spacelab, then by its own habitation module.

There have been other suggestions for building a permanent space station, or space construction base platform. Several studies have been aimed at reusing the

The large space station envisioned in the book, **Across the Space Frontier.** It was capable of holding more than 200 crew members. It will be many years before NASA will build a station of this size, but smaller, permanent manned satellites may be possible in the next few years. (Courtesy Viking Press)

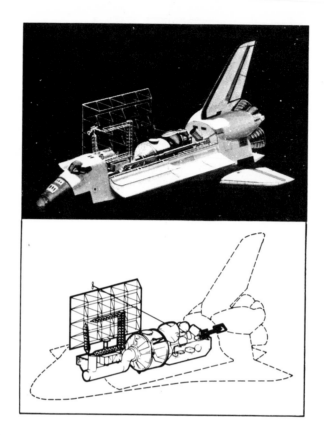

Two views of a large power module integrated into an orbiter carrying a spacelab. (Courtesy NASA)

only non-reused piece of the Space Transportation System, the giant external tank. Some proposals end with the external tank up in low-earth orbit, where several could be gathered in one orbital location by the teleoperator retrieval system or a similar propulsion device. Once in a central location, they could be stripped of their fuel fittings and made into habitability modules, one by one. After a space construction base with sufficient power was constructed, the empty external tanks would become the backbone of a giant construction, which would grow like a maze in space. This is another relatively inexpensive idea which could place a large space structure in low-earth orbit by the late 1980s, operating as a manned space station and initial construction base.

Looking beyond, there will eventually be what might be called a near-earth-orbit space community, composed of several different space habitats and facilities, such as space processing modules, construction platforms, energy plants, maintenance facilities, space observatories, and the old permanent manned wheel in space, as in the film, *2001*.

But near-earth orbit is not the only place where mankind will need permanent structures in space, nor is it the best place for several projects which can be envisioned. Many future space systems will be designed to operate in geosynchronous or Clarke orbit. In synchronous orbit, there is nearly continuous sunlight for energy systems and power for both the earth below and for the space facilities. There is also a continuous view of the same spot on the earth's surface.

Even as near-orbit facilities are being constructed in the 1980s and early 1990s, there will be activity high in the synchronous orbits, 23,500 miles (38,818.5 kilometers) above the earth. The first facilities in synchronous orbit will be carried to near-earth orbit by the space shuttle and to synchronous orbit by the interim upper stages. Later facilities for synchronous orbit, such as advanced communications antennas, will require additional and larger propulsion systems, as will manned construction platforms, habitability modules, and true space stations in the high orbits.

Not only will large propulsion devices be necessary for manned use of the high orbits, but there will be the need for an orbital transfer vehicle to transport large structures, heavy pieces of equipment, and large work forces. An OTV is needed to extend our capabilities from the low-altitude shuttle orbits to synchro-

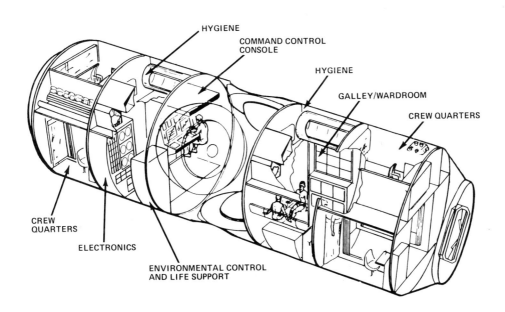

A habitability module would allow continuous living and working in space. (Courtesy NASA)

SPACE PLATFORM
POWER MODULE, MDA, ET

A space shuttle's external tanks could be used as a first step toward a space platform. Two thousand cubic feet (60.96 cubic meters) of the forward area would be equipped on the ground for crew habitation in low-earth orbit. An initial crew of three would inhabit the forward area of the tank and would construct additional living and working space in the remaining portion of the tank, as required. An external tank could provide as much as 19,000 cubic feet (5791.2 cubic meters) of room, useable as a small space station. (Courtesy NASA)

nous orbits. It might require assembly in low-earth orbit. It could be ferried up in two shuttle loads, or it might be flown up as an additional stage on the space shuttle, if the shuttle were made more powerful in order to handle the additional payload.

The orbital transfer vehicles could be assembled and fueled (or refueled) at the low-earth orbit space construction bases or at manned space stations. For unmanned transfer from low-earth orbit to synchronous orbit, the IUS would be used, or a solar-electric propulsion stage with a 100-day, one-way flight time. Large space structures, or assemblies, could be towed to synchronous orbit by manned OTVs, or sent there from earth orbit by improved and powerful IUS-type stages.

Once a strong foothold is obtained in synchronous orbit, the fantastic era of the industrialization and commercialization of space will begin to be a reality, instead of a future plan. With large space structures, platforms, construction bases in high-synchronous orbits, space will become the new technological frontier.

Beyond synchronous orbit, the manned station, or construction base, will still be the forward line of man's space expansion. Ultimately there may be large, complex, commercial space stations in low orbit and in synchronous orbit: Kraft Ehricke's space city, Stropolis, for example, and hotels—the low-orbit Hilton, the high-orbit Sheraton—and especially hospitals, using the zero-g of space to save patients who would have died on the gravity-enshrouded and polluted earth below.

The manned station may orbit the moon, as an orbiting lunar station, or a future variation of the manned lunar orbiting laboratory. When man steps away from the high-synchronous orbit, new vehicles will have to be developed and derived from the space shuttle. Large lift vehicles, or heavy lift vehicles, would be designed to take giant payloads to low-earth orbit, where large OTVs would transfer the payload to the manned lunar station. A single-stage chemical OTV can be easily designed to deliver a payload of *seventy tons* to a lunar station, and return nine tons to low-earth orbit. Such an OTV would be about 100 feet long (30 meters) and 26 feet (8 meters) in diameter, and use 380 tons of liquid oxygen/liquid hydrogen in a standard space-shuttle main engine.

Once near-earth orbit is developed and synchronous orbits have been exploited, the manned, lunar space station would be built as a preliminary to a manned lunar base and to the utilization of the resources of the moon for further construction and development in space.

There is even the possibility that heavy payloads may be lifted to near-earth orbit by a laser beam, instead of the conventional chemical rockets with which we are so familiar. A giant laser, which may be a future possibility, could lift a one-ton payload to a 100-mile (1000-kilometer) orbit using a gigawatt of power (equivalent to 1000 megawatts). Experiments have already been tried by hitting a projectile filled with an explosive gas with a high-energy laser beam. The gas ignites and thrusts the projectile forward. Theoretically, the concept might work for very large vehicles taking payloads up to low-earth orbit.

There is little in the planning for the future which is far-fetched. It follows a logical sequence, and it will only take time to make it past history instead of future hopes. Few people predicted artificial satellites by 1957, or their logical development into huge satellite systems which control our commercial and military communications, our management of resources, and our predictions of the weather. Few people really understood when the moon was reached that we might one day live there, perhaps on the other planets.

For those who see the logical extensions of the Space Transportation System, it is not difficult to imagine the sequence from the first spacelab to a free-flying manned laboratory. It is only a small step from there to a larger manned space station, and then only a step to the high synchronous orbits.

Prediction is slippery in these technological times, which have come upon us since the advent of the space age with *Sputnik* in 1957. But it is likely that there will be a space station circling the moon, from which men and women of different races and countries will look below at the stark lunar landscape of seas and craters, mountains and rills, before the first decade of the next century has passed.

On July 24, 1969, Trans World Airlines applied to the Civil Aeronautics Board for government permission to operate a passenger service ''between points in the United States and points on the moon.'' As far as anyone knows, they were quite serious.

The shuttle tomorrow is truly like laying the last spike on the transcontinental railroad, only much more so. And whether or not we're going to see it in the next 10 or 20 years, there are people (alive today) who will see manufacturing in space from moon materials or from asteroids.

Edmund G. "Jerry" Brown, governor of California, 1977

11

THE INDUSTRIALIZATION OF SPACE

"Space Factory 6 to orbiter. Space Factory 6 to Lafayette."

The shuttle comes in firing maneuvering engines, slows abruptly, and docks with the factory. "We show contact, Lafayette."

"Roger, six, we're down."

The giant cargo bay, filled with supplies from earth, opens. Men in extravehicular activity suits pour from the factory's airlocks into space to help with unloading. In five hours, the cargo is unloaded and a new cargo of precious electronics and special metals is roped and tied to hooks and latches along the sides of the space shuttle's cargo bay. The cargo bay doors close.

"Standby to clear, Lafayette."

"Standing-by."

"O.K., you're clear."

The shuttle drifts from the docking port by gentle firings from the vernier engines. As it moves away, the giant red letters on factory six stand out: RCA.

The "industrialization of space," which has also been called the Third Industrial Revolution, is often looked upon as a prospect for the future. It *is* a future **181** prospect, but it has also been going on for more than twenty years. When people speak of the industrial possibilities of space, space factories, and similar projects, they often ignore the current commercial uses of space.

One of the primary commercial uses of space at the present time is communications satellites. Commercial investment in Comsats totals well over a billion dollars. By the mid-1980s, worldwide commercial use totaling several billion dollars a year is projected. Relays now send telephone, radio, and television signals around the world. The service from satellites is far cheaper than long-distance cables, and they can link with any place on the earth which has suitable ground equipment.

The first attempt at this industrialization of space was the U.S. government's project SCORE or Signal Communication by Orbiting Relay Equipment, in December 1958. The first attempt at a communications satellite wasn't very impressive in view of later possibilities and potentials, but it was unique at the time. SCORE was simply an Atlas rocket in low-earth orbit carrying a tape recorder and radio equipment. It broadcast a prerecorded Christmas message to the earth and was also able to record signals from ground stations and retransmit them.

Transatlantic television arrived with *Telstar* in 1962, and was much improved by a more powerful satellite, *Relay*. Communications satellites were then lifted by more powerful booster rockets into the higher geosynchronous orbits where one spot on the earth would always be below the satellite, and three satellites could send and receive signals from nearly any point on the surface. *Syncom* went into synchronous orbit in 1963, followed by *Intelsat,* and by satellites of other countries who wanted satellite communications available. Most countries on earth are linked by *Intelsat,* less than twenty years after the space program was started with the successful launching by Russia of *Sputnik.*

Weather satellites are a good example of the service aspect of space industrialization. They have become indispensable for worldwide weather monitoring and long-range forecasting. The value of weather satellites to the earth's economy, which is dependent on the weather, is immeasurable. The benefits of advanced warning and planning are relatively invisible, but space meteorology has been said to earn more than fifty million dollars annually. Future space meteorology will include advanced operational systems with vastly superior and more sophisticated sensor systems. Eventually, it has been predicted, there will be potential modified control of the earth's weather through the use of large solar reflectors in space.

In this artist's view, the shuttle orbiter places an earth resources satellite in orbit. (Courtesy Rockwell International Space Division)

The earth resources satellites, called *Landsats,* have also furthered the commercialization of space. Cities, counties, states, reservations, countries, all can use the remote sensing instruments aboard the satellites to plan land use, develop resources, estimate reserves of materials, identify geological problems, uncover pollution, and research other areas. The earth resources satellites were developed from the early *Nimbus* weather satellites, which they resemble. These will be developed further and will become more sophisticated and more widely used by other countries.

Navigation satellites are relatively recent, and will be developed in the coming years. Satellite observation of sea and lake ice fields and icebergs has already extended shipping seasons and reduced shipping hazards to the point where the financial savings to the shipping industry far exceed the total research and development expenditure for the satellite services.

The navigation satellites, which continuously beam signals from circular orbits of 12,543 miles (20,186 kilometers), are a part of a larger program called the Global Positioning System. *NavStars 1, 2,* and *3* were launched in 1978 and became operational by November of that year. By 1985, there will be twenty-four *NavStar* satellites, which will give global navigation coverage under all weather conditions. So accurate are these satellites that positions on earth are pinpointed to an accuracy of 30 feet (9.144 meters).

But what we normally think of as industrialization has yet to come. Although the initial commercialization has been with earth resource, communication, and other types of satellites, the future will be most exciting in the relatively untapped areas. Space creates unique possibilities in a dozen different fields, and there are possibilities which will be discovered only as the industrialization process continues.

The vital characteristic of the environment of space is weightlessness, or zero-g. The absence of gravity makes possible manufacturing feats which are unfeasible on Earth—or on the surface of any other planet or moon. Some alloys of materials are immiscible—that is, incapable of mixing or blending—in the presence of gravity. Some products cannot be extracted from a mixture because of gravity, such as some vaccines and drugs.

A large number of products can be manufactured in space much better than they can on earth, either because of decreased production cost or greatly improved performance. Certain crystals for solid state electronic devices, for example, if manufactured in space, would be far superior to those made on earth and eventually much cheaper.

In space large structures can be made of extremely lightweight materials, because there is no gravity to distort the building materials. A spider web of a giant antenna can be built in orbit, which would immediately collapse of its own weight if it had been built on earth. Any space factory or construction base would be completely immune to earth hazards such as earthquakes, storms, hurricanes, and tornadoes.

The lack of gravity also allows complete mobility of work crews and construction equipment. There is no weather, nothing to hamper moving large structures from one place to another, nothing to stop work in the sense we understand it on earth. A three-shift, twenty-four hour day would not be unusual to space workers.

Research can easily be conducted into physical and biological processes in ways that now can only be imagined. A space hospital would have research facilities and an ability to investigate diseases that is unparalleled in human history. Space medicine may ultimately realize the fountain of youth—extended longevity, immunity to all disease, perfect health.

Space also offers a near-perfect vacuum, at least much better than that available on earth, and it is, of course, completely free. Large electronic devices in a vacuum can exist in the open, without having to be enclosed. A vacuum tube, although it is a bit old-fashioned, would not have to have the tube if it were used in space. There is unknown potential in the acoustic isolation possible when there is nothing to carry sound.

Possibly best of all, space is *huge*. It is an unlimited disposal area for heat, waste products, dangerous materials. There is no worry about storage room and practically no cost. Objects stored in orbit in space will stay there nearly indefinitely, without harm or visible deterioration.

Atmospheric pollution around the earth could be greatly reduced by moving polluting industries into space. There, waste heat is radiated harmlessly into the void. Dangerous waste products can be stored in orbit indefinitely, or shot into escape orbits to leave the vicinity of the earth. Waste products could even be sent away from the entire solar system—placed in parabolic orbit away from the sun's family, never to return.

Free and literally unlimited solar energy is available in space. All of our space satellites use free solar power. The vast potential for solar-energy power could solve the energy problems on earth in the coming century, as will be seen in a later chapter.

Finally, in the space environment perpetual motion is possible. In the complete lack of an atmosphere and other retarding forces, a factory set in motion will remain so, its operation bounded only by the reliability of the moving parts over a long period of time. Once a spacecraft or space platform is built and in orbit, the energy required to keep it in that location is extremely tiny. The *Lageos* spacecraft, which was orbited in the late 1970s, has an expected lifetime of eight million years.

The shuttle opens up all these immense possibilities for the industrialization of space. A few initial research experiments in materials processing were made

aboard Apollo, *Skylab,* and during the *Apollo-Soyuz* Test Project. The Russians have conducted fairly extensive preliminary experimentation aboard their space station, *Salyut.*

With the operational phase of the space shuttle and spacelab, there will be limited, small pilot plants performing various manufacturing processes. Future commercialization is as obscure to us now as the future of Manhattan might have been to the Pilgrims, or Detroit to the early explorers.

Certainly, judging from the past two decades, there will be hundreds of millions of dollars per year of commercial operations in space. From what glimpse we have today, the first commercial manufacturing ventures will probably be in selected electronics products and pharmaceuticals.

Two of the most important benefits of working and building in space will determine the extent to which the industrialization process continues: the safe, long-term availability of waste storage and the free supply of solar energy. What company can long resist free storage, an absence of federal waste regulations, and unlimited energy?

At first, only those facilities which can be deployed from the space shuttle's cargo bay will be used for manufacturing. The products will have to be small enough to be delivered to earth by the space shuttle. Later, facilities will be constructed which will remain in low-earth orbit and be serviced by the space shuttle. From these larger facilities will come the first major products of the industrialization of space. By the early 1990s, fully developed commercial facilities are expected to be in orbit and operated by private companies.

Among the processes which will be tried in space is levitation melting, or solidification of molten metals without benefit of a container. Levitation melting has several advantages over the methods employed on earth. Contamination of the materials by the container is eliminated allowing much higher purity. Alloys impossible on earth can be made, and perfect shapes can be easily and cheaply formed.

The melts will be heated by heating-coils and handled by the space shuttle orbiter's remote manipulator arm. By injecting gases into mixtures or metal melts, some extraordinary materials can be made: steel as light as balsa wood, mixtures of glass and steel impossible to achieve on earth.

The size of earth-formed crystals, used in electronics, is severely limited, and they are often flawed by microcracks and contaminants. They are also relatively expensive. Crystals made in space would not be contaminated by

One of the first experiments toward eventual industrialization of space will be the solar array wing. The experiment will be flight-tested aboard an orbiter in 1980. The wing is 105 feet (32 meters) long and 13½ feet (4 meters) wide. It will be extended and retracted from the space shuttle's cargo bay to test structural and dynamic characteristics and electrical performance. With eighty-two solar cell modules, it could be used to power extended spacelab missions. (Courtesy NASA)

containers. Internally they would be uniformly distributed, and they can be extremely large in the zero-g. A space factory for manufacturing silicon crystals would be completely automated, operating on energy provided by solar arrays. The shuttle orbiter would periodically visit the space factory, delivering raw materials and picking up finished crystals for earth.

According to studies conducted by General Electric, the break-even point for space-made, flawless crystals would be about twelve years. Meanwhile, the research and development into crystals and allied fields would lead to new types of bubble memory components for computers and to additional capabilities in microminiaturization, especially useful in medicine.

Glass made in space would not be contaminated by contact with a mold. Contamination of glass on earth causes crystallization. Also, in space, higher manufacturing temperatures can be used. Glass produced above the earth would feature new materials to improve the optical design of telescopes and microscopes. Lenses made from space glass would make the average camera buff hysterically happy.

In the future, space medicine would be a wide-open field. The possibilities are endless. For example, electrophoresis would be a powerful analytical tool in space. Electrophoresis is the migration of fine particles of a solid suspended in liquid to the anode or cathode when an electrical field is applied to the suspension. Isozymes could be separated by electrophoresis and then studied. Of the two thousand enzymes which control our metabolic processes, at least one hundred are mixtures of isozymes. If the isozyme groups could be isolated and studied, we could predict body disorders and imbalances much earlier than by currently available diagnostic techniques.

To medicine, one major benefit from space would be the production of high-quality urokinase. Urokinase is a catalytic substance created by a small group of cells in the kidneys. It prevents blood clotting, and it is far more precious than gold. A ton of urine must be processed to get one dose of urokinase, costing $1200. Electrophoresis could be used to produce the substance in quantity, at a relatively cheap price.

A future development of space will be tourism, which should delight those people who would never have a chance to be an astronaut, mission specialist, or payload specialist aboard the space shuttle. For those people who willingly pay $2000 to fly the Concorde, or five times that much to take a trip around the world, excursions to the world of near-earth orbit may be available during the next two decades.

If the commercialization of space extends to tourism, then the businesses and industries which cater to tourism will inevitably follow. It is not beyond belief or possibility that the equivalent of a *near-earth Hilton* will be built in space before the end of this century, certainly by the middle of the next. Space tourism is only a matter of time, and the conquest of space itself has been a very short interval. It

Shuttle docking with a permanently orbiting space factory, using the international docking system originally developed during the Apollo-Soyuz flights. After docking, personnel could be transferred without spacesuits to the orbital facility. (Courtesy Rockwell International Space Division)

An artist's view of the assembly of large space structures. The materials have been taken into space by the orbiter. (Courtesy Rockwell International Space Division)

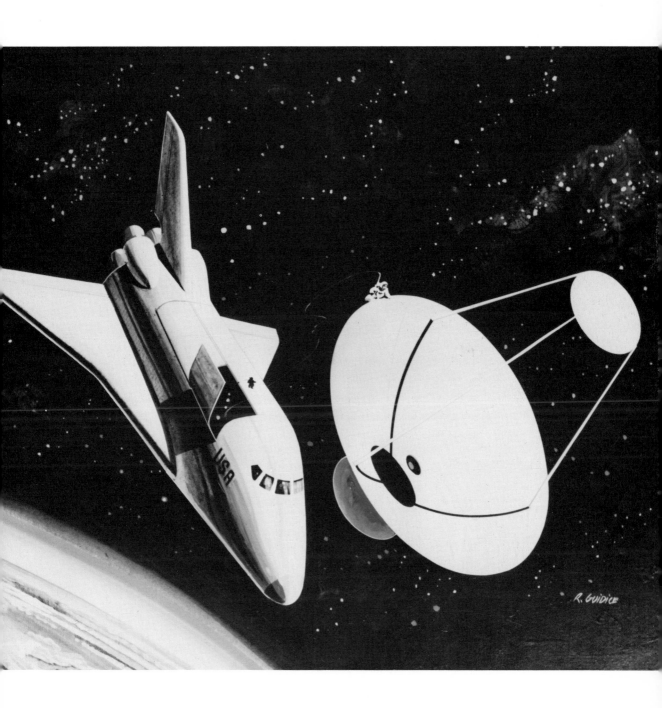

was just slightly more than seventy-five years ago that the Wright Brothers made the first flight in a powered, heavier-than-air craft. It lasted for twelve seconds and covered a distance *two feet shorter* than the length of the present space shuttle orbiter. The next seventy-five years are likely to bring an equal jump.

In addition to plans for placing complex and more sophisticated satellites into orbit in the future, there are also studies aimed at a new satellite system for use with the space shuttle—the Shuttle/Tethered Satellite system. A satellite payload would be tethered from the shuttle's cargo bay, either towards or away from the earth, depending on the application, at distances up to 62 miles (100 kilometers) from the orbiter. The tethered satellite system would be capable of multiple round-trip missions, and could be used to commercially map the earth's magnetic field and to explore the upper atmosphere. The system could aid the coming industrialization of space by transfering cargoes and assisting in the erection of large space structures.

One of the keys to future industrialization of space is the 25-kilowatt, and larger, power modules mentioned in the last chapter. These, plus continuously manned habitability modules, or space stations, will open the way for hospitals, tourist hotels, and space factories of all kinds. Before it is completed, the first inhabited station will be used as a construction base for testing future high-level space technology.

For the initial period of space development, relatively small structures will be needed—165 feet (50 meters) in diameter. By the period 1985–92, it is expected that structures as large as 200 to 300 feet (60 to 90 meters) will be needed. For the far future, and for giant solar energy needs, space structures several miles in diameter will be required.

Part of the planning for space includes very large space structures of three kinds: booms, dishes, and planar surfaces. Large dishes will be used for earthside and deep-space communications, earth observations, astronomical exploration, a search for extra-terrestrial intelligence or SETI, and power generation and transmission. Booms will be used for locating positions and for low-frequency communication. Planar surfaces will be extensively used for power transmission, communication and facsimile transmission, and space radar.

An earth-viewing dish with a diameter of 160 to 640 feet (50 to 200 meters) will provide the first useful soil-moisture data for planning and managing agriculture. A deep-space relay dish would replace existing earth-based systems for communications with interplanetary unmanned probes, such as the *Galileo* to Jupiter and others destined for the ends of the solar system. It could be used to communicate with roving vehicles on Mars and high-flying robot vehicles in the atmosphere of Venus.

Opposite: Deliveries are made to the construction site of a solar satellite. A shuttle carrying workers docks at the assembly bay of a solar satellite while, at the bottom, a heavy lift vehicle delivers construction materials. (Courtesy Boeing)

Left: A future heavy lift space freighter may deliver up to one million pounds to an orbiting solar power satellite construction base. (Courtesy Boeing)

Below left: Construction of a thermal engine power satellite. The sun's rays would be reflected into the furnace cavity at center by the extensive field of solar reflectors. (Courtesy Boeing)

Below right: A crew member photographs the repair of a jammed beam-building machine. (Courtesy NASA)

Early communications projects for large space structures to be deployed by shuttle are a bootlace array and a parabolic dish. Experiments and use of these early devices would precede the construction of a multibeam communications antenna. A multibeam antenna 200 feet (60 meters) in diameter, weighing 60,000 pounds (27,000 kilograms) would be powered by solar cells and could provide communication services on earth for two and a half million individual users.

An even grander version is a personal communications space antenna. If built in space, such a communications network would completely revolutionize the telephone call. A call to Europe or next door would be completed in space. Correspondents could be dialed directly on a small wrist telephone or hand-held communicator. It would not even be necessary to know the location of the person being called. Connection would search out the wrist communicator anywhere on earth. And the interference and background noise would be about the same, or a little less, than that which comes from a normal FM radio receiver.

A personal communications satellite would not be a small antenna; it would contain nearly 70,000 square feet (21,000 square meters) of solar collector area to provide the energy for three 83-foot (25 meters) separate antennas, each with a huge capacity.

The cost for personal communication would be much lower than current long-distance rates, once the original investment in the satellite was recovered. A call across America would be as inexpensive as a call across town. A personal communications satellite would make a Wats line available to anyone for the price of a home phone today.

Another exciting possibility in personal communications is the electronic mail satellite. The United States Post Office's new motto could be "through rain, sleet, and snow, and meteorites." The huge, spidery electronic mail assembly would be fabricated on the ground and transported to low-earth orbit by the space shuttle. With the orbiter as a construction base, the mail antenna would be assembled and pushed into its operational orbit by low-thrust propulsion systems such as the teleoperator retrieval system.

The electronic mail antenna would orbit the earth at an altitude of 345 miles (552 kilometers) on a platform constructed entirely of identical struts and unions. It would take 2650 double struts and 619 nine-point unions to build the platforms 900 feet (270 meters) in diameter. The space mail station could be connected with 1274 postal centers and 30,000 post offices, and could carry one hundred billion pieces of mail per year which would save one billion dollars a year at current delivery costs. A letter would be delivered electronically by the satellite, using facsimile scanning. Eventually, the mail service would be cheaper and much more reliable than that of today. And it would be considerably faster. Other versions of the electronic mail system would place the satellite in a higher, geosynchronous orbit to serve the continental United States.

Opposite: Interior of a cylindrical colony of the future. (Courtesy NASA)

A concept for a very large system antenna. There are two small "feed" spacecraft, relay satellite, and a radio-frequency interference shield. The system and antenna could be used for a search for extraterrestrial intelligence. It would be located in geosynchronous orbit or beyond. (Courtesy NASA)

Once the commercialization of space has begun in the near-earth orbits, it will gradually continue up to the geosynchronous orbits. One example of a proposal for the future is the geostationary space platform. The platform would be a collection of antennas performing a wide range of communication, detection, and control functions for the public. Within its possibilities are the electronic mail service, police communication, personal communication, disaster control, earthquake detection and prevention, water availability indication, vehicle speed control, and burglar alarm and intrusion detection. All of the services would eventually be at a vastly reduced cost, relative to today.

A gigantic array of antennas could be placed in geosynchronous orbit to relay electronic mail. The huge system could be serviced by space shuttle and would make mail delivery over most of the United States a matter of simultaneous transmission. (Courtesy NASA)

All of the various elements of the geostationary space platform would use the same power systems, and station keeping, altitude control, tracking, and control. The station would be resupplied and maintained with a single rendezvous and docking adaption. Later, the facility could be continually manned by technicians, supplied by the successors to the Space Transportation System.

The use of space for material processing and manufacturing platforms will follow the same pattern as the development of space stations outlined in the last chapter. Spacelab will have extended missions with a power module, then habitability modules especially for testing and pilot plants will be built. Eventually, there will be manned factories in near-earth orbit, and some will move outward to the higher geosynchronous orbits.

The tip of the space-use iceberg is hardly visible. By the time the whole is exposed, the world which it will create for mankind might not be recognizable to the people on earth today. But if the preliminary indications are to be believed, it will be a world so sophisticated in the use of space technology that most, if not all, of mankind's problems for the next few centuries will be controlled or completely solved.

Electric power on earth between now and the end of this century is a trillion dollar business. Even a small percentage of that market would dwarf total expenditures on space research and development since 1958.

John H. Disher, director of Advanced Programs, Office of Space Flight, NASA, 1976

12

SOLAR POWER SATELLITES

Power! Energy! That was what the earth wanted and was running out of. But the space shuttle and stations changed all that. The cargo bay of the shuttle housed the extending solar panels which performed the first large-scale experiments in solar power generation. Systems which earth satellites and spacecraft had used for twenty years were refined, improved, and made huge.

When the construction bases had been begun and the space stations manned, automatic beam builders and the highly paid crews began the solar structures. They rode the beams above the earth, a hundred miles high, sometimes two hundred, while the space shuttle brought more materials and men to the construction bases. When the first segment had been built, it was tested. It put out more energy at a better conversion rate than had been thought possible only a few years before.

Spidery structure by spidery structure, the first solar satellite took shape and was moved from the construction orbit to an orbit well away from the growing commerce in space. It powered an earth city, along with help from the ground power stations.

197

When ten years had gone by, the orbits were full of structures, and the land below was being freed by the sun. For the second time in the history of mankind, the elements of the universe had been broken down and made useful. Once when the atom had been shattered; now solar power stations took the free energy of a constant thermonuclear reaction known as the sun and converted it to a source of raw power for the people of earth.

One of the major events in the future development of space is the construction of a solar power satellite. In scope and commitment it would rival the building of the pyramids, the raising of the giant stone gods on Easter Island, the assembling of Stonehenge. Its cost would rival all wars, and be sufficiently high to bankrupt smaller countries. As a sustained technological achievement it would make the lunar landing nearly insignificant. The purpose would be to provide the raw electrical power to supply the United States' energy requirements for the next centuries with solar energy harnessed in orbit and directed to the earth for distribution.

A solar power satellite, or a satellite power system, was first proposed in 1968 by Dr. Peter Glaser, vice president of Engineering Sciences at Arthur D. Little, Inc. His basic concept was giant satellites covered with solar panels which would collect sunlight twenty-four hours a day. From a high orbit of approximately 22,000 miles (35,404 kilometers) above the earth, the solar power satellite would beam the sun's energy to the surface in the form of microwaves. The microwaves would be converted to electrical energy on the ground. The output of a 72-square-mile (115.2-square-kilometer) area would be enough to power at least one million homes. Some early plans called for as many as one hundred of the giant power satellites circling the earth by the year 2025.

The design of solar power satellites has, in the years since the idea was first introduced, evolved into many different forms with designs contributed by a dozen different companies. Design of the antenna and the ground receiving apparatus varies considerably. But one idea is in common in all the designs: the satellites could not possibly be built on the earth. The construction of a solar power satellite system would require perhaps four hundred space construction workers and a base construction platform as large as Manhattan island.

Estimates of the cost of a solar power satellite system are staggering. Development costs range from $60 billion to $80 billion. To orbit a sufficient number of the satellites to supply the United States' energy needs, the costs have been estimated at between $1 trillion and $2.5 trillion. No other planned or contemplated space project has made as many friends, or as many enemies, as the solar power satellite proposal, which is generally known as Sunsat.

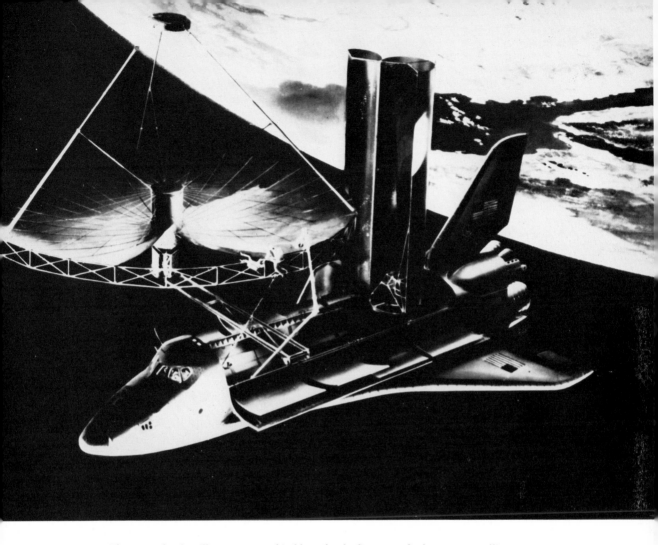

The space shuttle will serve as an orbital base for the first tests of solar-power satellite systems on a small scale. A test system, shown here, would be a first step toward larger solar-power systems in orbit. (Courtesy NASA)

Opponents of Sunsat point out that the sun shines on earth, too, and there is no reason to spend billions of dollars to reach the sun in space. Other opponents are so stunned by the enormity of the concept and the frightening costs that they turn away from it in horror. Still others see it as a deliberate attempt to sell a program to the public in the midst of a genuine energy crisis in order to prop up the sagging aerospace industry and increase the size of NASA.

Supporters of a solar power satellite system counter adverse reactions by showing that sunlight striking the earth in Arizona—the best location for ground solar collectors—is only available seventeen percent of the time, while the United States' nationwide average is much lower, about six percent. In space, sunlight would be available ninety-nine percent of the time. The other one percent would not be available because of occasional passes of the satellite through the earth's shadow. A power collector in space would collect six times as much as one in Arizona, and seventeen times as much as one receiving the national average of useful sunlight.

The cost of a Sunsat, everyone agrees, is enormous, but so are the potential profits. If the power from a photovoltaic satellite were sold at cost of 27 cents per kilowatt hour, the power output would be equivalent to a $2 billion revenue annually. Looked at another way, a square kilometer of space would yield more than $15 million a year, even when its output is sold at cost.

As for the gigantic investment involved, admirers of the Sunsat point out that present electrical power transmission is inefficient and very costly. There are now more than 400,000 miles (643,720 kilometers) of high-voltage power lines in the United States, and the rights-of-way for these tie up 11,000 square miles (17,600 kilometers) of real estate. And the present system will not be adequate for the future. The capital investment of a solar power satellite system, according to several estimates, appears to be about $1500 to $2500 *per kilowatt*. Although the capital investment is enormous, it is still about the same as that required for a nuclear power plant which could be built in the same time period.

The solar power satellite system does not require future technology. It could be built with existing technology—although in the near future it might be considerably cheaper as new space construction techniques are developed, using the space shuttle. Current estimates indicate that the first Sunsat could be in orbit by 1995, and the system could be operating and providing the entire United States with power by the first decade of the twenty-first century.

The Sunsats would be placed in geosynchronous orbit around the earth. The solar collector face of the satellite would face the sun, while the transmitting antenna would always face one spot on the earth's surface. The Sunsat's solar collectors would be composed of millions of photovoltaic cells which would convert the solar energy directly into electricity. The electrical energy would then be fed to a microwave antenna, which would rotate once a day with respect to the solar collectors. The electrical energy, converted to microwaves, would be beamed to the earth's surface to a receiving station, called a rectenna, where they

The central core of an incomplete solar satellite is shown in this artist's view, looking toward the sun. A solar-power satellite would be placed in geosynchronous orbit. (Courtesy Johnson Space Center and NASA)

Another version of a solar-power satellite. The system would be placed in orbit to collect sunlight, which would be converted to energy and transmitted to earth by microwave beam. The beam would be converted to electrical energy at earth-based stations. The solar-power satellites would be very large structures, fabricated from materials delivered to low-earth orbit by the space shuttle. Some solar power designs, such as the one shown, could provide electrical power on the order of 5000 to 10,000 megawatts. (Courtesy NASA)

would be converted to direct current. Simple dipole rectifiers on earth can convert microwaves to direct current, DC, with an efficiency of ninety percent.

In 1975 in the Mojave Desert at Goldstone, California, microwaves were converted directly into electricity with an efficiency of eighty-five percent. Although the distance was only one mile, the principle is the same as for a solar power satellite. Utility power stations, in contrast, are only about thirty-five percent efficient.

The photovoltaic cells, which a Sunsat would use, are currently very expensive to manufacture, contributing to the high cost estimates for a Sunsat.

A ground-based microwave receiving station for a solar-power satellite system. The receiving station would be near the consumer, largely independent of weather conditions. (Courtesy Raytheon Company and NASA)

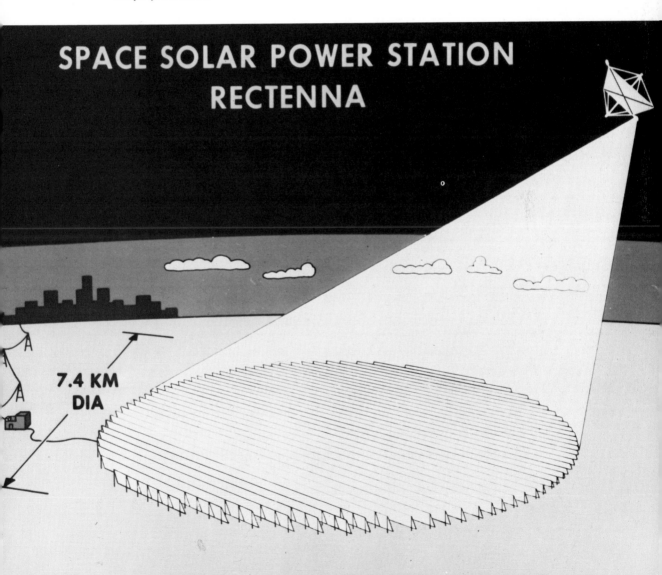

However, recent advances in solar cells using amorphous, uncrystallized, materials instead of silicon indicate that new forms of solar cells will be available by 1981 which can produce electricity, at least on earth, at a cost comparable to power from a conventional coal or nuclear plant.

For the cost of a solar power satellite to be practical, the photovoltaic cells would have to be cheap and able to inexpensively blanket square mile after square mile of satellite structure. The new amorphous material may be the answer to part

The first structural member of a giant solar-power satellite is being fabricated in this artist's concept. (Courtesy Johnson Space Center and NASA)

of the cost problem of a satellite system. With the new cells, the satellites would be much smaller or fewer in number than if they were composed of pure, crystal silicon cells.

With crystalline cells, the expense comes from the necessary purity, and the cost of cutting and growing the crystals. Crystals from space factories are one answer to the problem, but cells of other materials are a quicker one. The new, amorphous-material solar cells convert energy with an efficiency of about ten

A beam builder at work in near-earth orbit. Beam builders would fabricate a large structure as a development step toward future space solar-power systems. In this illustration, the beam builder is fabricating a final longitudinal member. (Courtesy NASA)

percent and have about a twenty-year lifetime. Silicon cells are slightly more efficient at twelve to fourteen percent.

One of the most interesting aspects of the new solar cells is the proportion of the solar spectrum which would be converted to electricity. By adding fluorine and reducing the hydrogen content of the materials, the new solar cells alloy can convert a wider band of the solar spectrum into energy, adding to the efficiency of a solar power satellite system.

Estimates of the necessary number of solar power satellites vary, partly because the photovoltaic cell industry is in its infancy, partly because the whole concept of the Sunsat is changing as space construction techniques are thought about and discovered.

A Sunsat using existing solar cell technology—silicon cells, not the newer amorphous alloy—with an antenna 7 kilometers in diameter would service a city the size of New York with all its power needs—10 million watts. The satellite would have a surface of about 100 square kilometers. Twenty-five satellites of this size would provide the electrical needs of the United States today.

A study of the solar power satellite has been funded by NASA for four years. Congress has allocated $15.6 million for research into the concept. Additional funds of $25 million have been requested for study. One aerospace contractor, the Boeing Company, which has studied the concept extensively, believes that over the next five years, more than $3 billion will be allocated by Congress for the Sunsat concept study and the construction of a prototype satellite in space, using the space shuttle technology.

The Sunsat program is now in the research and development stage. When preliminary studies have been completed and perhaps a prototype satellite, or satellites, has been built, then the decision will be made whether or not to build a Sunsat system. The huge cost of a Sunsat system may not ultimately be the determining factor. Congressional testimony has said America needs three hundred additional light-water, nuclear-reactor power plants in this century, with each plant costing $2 billion in today's dollars. For a $600-billion investment, the solar power satellite system begins to seem much more feasible. Add the dream of clean, free power without waste, with most of the power industry cluttering up space, instead of the earth below.

Opposition has focused on the potential dangers of the microwave beam of a Sunsat to life on earth and to the atmosphere. Some experts believe the heat carried by the beams would do extensive damage to the ionosphere and other parts of the upper air. Other experts say that the heat from a microwave beam would be so diffused that it would have little or no effect on the earth's atmosphere.

Cargo and supplies are delivered at a completed solar-power satellite by shuttles or other highly developed cargo vehicles. Solar-power satellites may be the main energy source for the coming century. (Courtesy NASA)

Other critics of the Sunsat have claimed that the microwave beam would be extremely dangerous to nearby airplanes, and that birds would be cooked if they flew into the beam path. Because of the speed of airplanes and the relative low density of the microwave beam, other scientists have said that there will be no potential danger to aircraft from a Sunsat beam. Calculations from some sources have shown that the density of the microwave beam would be about one-fifth the density of sunlight and therefore hardly damaging to anything. On the edge of the microwave beam, the danger from exposure to microwaves might be as low as one hundred times below the present United States limit.

Could it be built soon? The answer, as it often is in space technology at the present time, is yes, if the financial commitment is made. The technology is available or could be developed quickly.

The first step toward building a solar power satellite will be developing techniques for space construction and for making and erecting structures built of beams. In 1983–84, employing the space shuttle orbiter as a construction base, the automated beam builder, used for the first time, will test some of the techniques which might later be applied to the construction of solar power satellites and other large space structures described in the chapter on space industrialization.

These techniques will be employed in the construction of the first, prototype solar power satellite. If the prototype shows that a Sunsat system is the energy producer needed for the future, and is as practical or more practical than other forms, then one of the most grand dreams of mankind will start to be realized in space.

A demonstrator automated beam builder exists currently. It can produce 1-meter aluminum beams from .016-inch, flat plate material. First used on the ground in early 1978, it was to be delivered to NASA near the end of that year. It was designed to be able to produce beams as long as 532 feet (162 meters) without reloading. With another design, the ABB will handle beams 984 feet (300 meters) long without reloading; a future design, 1500 meters. The ABB can be adapted to commercial roll forming, welding, and other forms of automated fabrication and assembly.

The space construction devices that will supersede the automated beam builder will be the products of a new and developing technology. As larger and larger structures are built, the means to build the solar power satellite system will become more practical and ultimately cheaper.

The second stage of a heavy-lift space freighter glides to a landing after delivering an almost one-million pound payload to an orbiting solar power construction base. This possible successor to shuttle would also have a reuseable first stage. (Courtesy Boeing)

The space shuttle is hopelessly outclassed as a cargo carrier for building a solar power satellite system. New launch vehicles would have to be designed and built to carry the tons and tons of materials into orbit. One designed giant cargo carrier looks like a very oversize Mercury capsule. At lift-off it would weigh 12,000 tons and be capable of carrying more than 500,000 pounds of payload into orbit. This cargo carrier, a heavy-lift vehicle with twenty engines each equal in thrust to four Saturn V rockets, would carry construction materials to an assembly platform or construction base with facilities for hundreds of astronaut-construction workers.

After its completion in near-earth orbit, a Sunsat would slowly accelerate toward its home in the geosynchronous orbit on solar powered electric engines. It would arrive after a few weeks perhaps carrying a group of workers to make sure everything functions correctly. A giant Sunsat, slowly drifting up to a high orbit with the construction workers aboard, recalls the story of the Brick Moon in chapter two. Perhaps as in the story, the construction workers would have their families, or at least their spouses, along for the trip to high orbit.

13

BEYOND THE SPACE SHUTTLE

The ship hung in orbit a thousand miles above the earth. Nearby, two other ships hung equally motionless in the black velvet.

Slowly they began to move away from the orbit. Soon they had passed the speed which would have taken them to the moon. They drew onward into space gathering speed, and swung out in a wide orbit which had an intersection point with the orbit of the fourth planet from the sun. The first manned Mars expedition was underway from High-Orbit Construction Base 3. It would report back in a few years with pieces of Mars, samples of the air, small boxes of the soil, and perhaps a scrape of some exotic biology as yet undiscovered by the three unmanned missions which had landed on the planet.

The century had already turned. The space shuttles were different now, but still being a shuttle, came and went from orbit to orbit, surface to space. The cities of earth would not have been recognized by those who had designed the first space shuttle in the ending years of the 20th century. It was a different time now with a different technology. Soon man would have a mining base on Mars, and eventually cities as they already were on the moon. It would be time, soon, to look far outward toward the nearest star and dream the old dreams.

211

Space over the next few decades will be a complex world, if the opportunities which the space shuttle will open up are seized quickly and pursued with a heavy commitment from both government and private industry.

There will be extensive use of the lower near-earth orbits. Communications and public service platforms could be in the geosynchronous orbits. Giant power satellites could provide the earth with energy. Factories in space could produce new products impossible below. And thousands of people could be living in manned space stations, in habitability modules on construction platforms, and in technical support facilities. All of this is possible within twenty to forty years, with current or easily foreseeable technology.

Beyond the geosynchronous orbits, there will be increasing use of orbits around the moon, in preparation for a manned, permanent, lunar base. One of the first efforts toward a return to the moon may be a revival of the concept of an unmanned, lunar orbiter, launched this time from the cargo bay of the space shuttle. The lunar orbiter's more sophisticated sensors, developed from the earth satellites, would search out mineral deposits and other commercial resources, in addition to completing the mapping of the lunar surface.

Before the moon can be used as a manned base, it will be necessary to build two new elements of the Space Transportation System: a manned, lunar space station and a suitable lunar lander.

An orbiting lunar station would be a six-to-eight-man station operating in a lunar polar orbit at an altitude of 68.8 miles (111 kilometers). Because of gravitational disturbances to a low lunar orbit, the station would maintain position with small power modules. Originally studied under NASA contract in 1971 by North American Rockwell, this would be a manned version of the long proposed lunar polar orbiter, which has appeared on NASA budgets for several years and has never quite managed to pass Congress.

The lunar station would be supported by supply flights of highly developed orbital transfer vehicles, either from staging areas in the geosynchronous orbits or stockpiles brought up to near-earth orbits by the space shuttle. The station would serve several functions: a base camp for lunar landers, a staging point for rescue from the lunar surface, and an observation and communications center for lunar surface explorations. In addition, the lunar station could select a suitable site for the first permanent lunar base.

A base on the moon would not be simply a scientific or technological delight. The development of its minerals and resources is an important step toward getting the necessary materials to continue expansion into space. It will ultimately be necessary that the economic resources for the development of space come directly from space. Earth's resources are already reaching a foreseeable limit, and the economics of transferring materials from its surface to space are frightening. Apart from the impossible cost itself is the near impossibility of receiving political approval for expending earth's dwindling resources in an expansion into space.

A nuclear orbital transfer vehicle separates after placing a manned lunar surface payload in orbit. This vehicle will extend man's orbital operations from high-earth orbit to lunar orbit. It could be used to build a manned orbiting lunar station and to ferry supplies, with an appropriate lander, to the surface of the moon where a base would be constructed. (Courtesy Rockwell International Space Division and NASA)

A manned base near the lunar south pole. In this artist's conception, an Antarctic-style operations hut with processors and power stations is in the background. In the right foreground is an astronaut's landing capsule. (Courtesy NASA)

A processing station for moon mining of water could be dropped from a lunar polar orbiter near the moon's south pole. The processing station would take soil samples from a robot rover and try to produce water and oxygen. (Courtesy NASA)

Studies by NASA in 1976 and 1977 concluded that lunar mining and space processing could begin by 1992 with an investment of approximately $60 billion, in 1978 dollars—slightly more expensive than the Apollo program.

The lunar base would be formed of module stations ferried by lunar landers from the orbiting station. Several modular space stations have been designed, each large enough for a crew of twelve.

The modules could be launched to low-earth orbit by the space shuttle and then carried to the vicinity of the lunar orbiting station by an orbital transfer vehicle. Once the landers had ferried the lunar base down to the surface, the crew and other construction astronauts from the orbiting station could build the lunar base. Unprocessed lunar materials would be used to make a protective covering over the base module.

At first, the lunar base would need to be totally supplied from the earth, or at least from the space stations. But later it could become more self-supporting. Air and water could be recycled, and lunar resources could provide materials for construction, soil for small farms, and fuel for nuclear power reactors.

The moon has abundant resources to support a future base: oxygen for life support and rocket propulsion; metals such as aluminum, magnesium, iron, and titanium for structures and also for rocket propulsion; raw materials for ceramics and glasses; silicon for photovoltaic solar power devices; and thorium to fuel nuclear breeder-reactors. Almost all of the concepts for lunar base self-sufficiency include raw materials of the moon. One of the largest of the resources of the moon is the solar energy which shines down on the surface during the long lunar day. A day on the moon is almost two earth weeks long.

To quickly lessen the lunar colony's dependence on the earth or the space stations for supply, the lunar base could introduce early a system of lunar agriculture combining physical and chemical processing, food plant growth, and recycling wastes with soil bacteria. Ultimately, a lunar colony might have to depend on earth only for some nutrients needed for food growth, for luxuries, of which a lunar base would have very few, and for hydrogen.

The moon, unfortunately, has very little hydrogen. The small quantity of hydrogen imbedded in the lunar soil comes from the solar wind. The lack of sufficient hydrogen to produce water and high-energy rocket propellants will be severe at the lunar base, and it is likely that the most critical import from earth or space-station staging areas will be hydrogen.

The lunar base has been considered as an alternative first step to the giant solar power stations for use on earth. If the power stations were built on the moon by beam builders from processed lunar materials, the energy to launch them from lunar to earth-geosynchronous orbit would be only one-twentieth of that required to launch portions of the power satellites from the surface of the earth. It might be cheaper, in the long run, to mine, refine, manufacture, and construct the solar power satellites on the moon and then send them to high-earth orbit.

Later in the development of moon mining, minerals could be mined from a lunar base. (Courtesy NASA)

The 1971 lunar rover. This 400-pound moon vehicle would be redesigned for the future, but the use of something like it on the moon would be invaluable. The 1971 rover could go 10 miles per hour. (Courtesy NASA)

A lunar base or colony which could manufacture solar power stations would be worth tens of billions of dollars each year. If the solar power satellites become a necessity for earth, then the construction of a lunar base would be economically practical, paying for itself by manufacturing power.

It is also possible for a lunar base to be used only to mine minerals needed to build structures in space. Initially, the lunar base would also process the ores. Eventually, the lumps of ore would be launched from the moon to an orbital manufacturing facility in geosynchronous earth orbit, or possibly in orbit around the moon.

Since the lack of hydrogen for rocket fuel would be a serious drawback to conventional rocket propulsion, the lunar base would launch ore with a mass driver. Superconducting magnets would lift a vehicle or load above a track stretched several miles along the lunar surface. Electric motors would provide thrust in the straight line. A nuclear generating plant would provide the power. With careful control of the direction and speed of the ore-launcher, the lunar metals and materials would arrive at the orbiting manufacturing facility, where they would be received automatically.

A lunar base is a logical step in mankind's expansion into space. It may become much more than a future possibility long before most people would expect. The earth's resources are dwindling, and they cannot be replaced. Thorium and uranium, predicted to be extremely useful and unbelievably expensive in the future, are already hard to obtain on earth. Although it would take as much energy to mine them on the moon—where we know from the Apollo missions that they are present—the possibility of free energy in the form of solar power might make a lunar mining facility profitable in the not-so-distant future.

Some critics of the space program in general regard a lunar base with a certain fascinated horror. In the midst of what they usually regard as a waste of time exploring and developing space for vested interests, there are proposals to spend billions of dollars to land again on the moon and actually live there.

While it is true that estimates for constructing a lunar base are in the area of $1 billion (1975 dollars), it is also true that calculations of the energy consumption on earth necessary to provide all the world's people with a degree of affluence approaching that of the United States would raise the earth's average temperature by as much as ten degrees. Only a half degree change in the average temperature costs billions of dollars in food production.

For the near future, the choice will be severely cutting back for generations all hopes of maintaining the present standard of living, or moving into space and using its raw materials and resources to meet the earth's new demands. If that is ultimately the question, the decision, considering the history of mankind, will be to develop space regardless of the initial cost.

Although some studies have advocated a permanent lunar base as early as 1998, it is unlikely that it will come that soon. But the world of 2030, only a little

A distant future lunar rover might be developed from this mobile geological laboratory, shown in a volcanic crater in southern California. (Courtesy NASA)

over fifty years from now, should be a world of traffic between the earth and the moon; of lunar mining with giant power satellites supplying both the earth and the lunar mines; of cities on the moon; and of huge space stations manned with thousands of technicians to keep all the space technology progressing.

Once the lunar base, with its mining facilities and minerals resources, has been developed, the possibilities for the uses of space between the earth and the moon are only limited by ideas. One of the ideas—based on lunar mining for materials, space technology for construction, and solar power satellites for energy—is the space colony.

The concept of a space colony is relatively new and is quite different from the concept of a space station, a giant hotel complex, or a space city. It was originally proposed in a *Physics Today* article by a Princeton professor, Dr. G.K. O'Neill, in September 1974. He later expanded the concept into a book, *The High Frontier*.

Since the introduction of the idea, increasing attention has been given to space colonies and their possibilities. T.A. Heppenheimer, a planetary scientist with a doctorate in aerospace engineering, developed the concept, complete with the interior design of a colony, in his book, *Colonies in Space,* published in 1977.

One approach to a space colony, envisioned by Dr. O'Neill, is large colonies made up of pairs of cylinders, with each cylinder 18.6 miles (30 kilometers) long by 3.72 miles (6 kilometers). The cylinders would rotate at a rate which would give the equivalent of one earth gravity on the inner surfaces. One end of the cylinders would be pointed toward the sun; on the outside, there would be large mirrors to reflect sunlight into the interior of the colony. The mirrors would be moved to create a twenty-four hour, day-and-night cycle reminiscent of earth.

At the start, a space colony would be small, perhaps a cylinder 0.75 miles (1 kilometer) in length and 1300 feet (400 meters) in diameter. Space construction men and women, up to ten thousand people, would use the smaller colony as a base of operations while building the large version. Raw materials would come from the moon, and the energy for building and maintaining the structure would be generated by sunlight.

The space colony would be located at one of the Lagrangian points (probably L_5) in a stable orbit. Stable orbits around the moon are not plentiful because of the constant pull of both the earth's and the sun's gravities, as well as the lunar attraction. The Lagrangian points are locations where a spacecraft, a space station, or a space colony, can be located so that it will always remain in a fixed position in relation to the earth and the moon.

There are five Lagrangian points, L_1 through L_5, named for the French mathematician who first discovered them in 1772. L_1, L_2, and L_3 are on a line between the earth and moon and are relatively unstable. If a spacecraft placed at one of these points moved slightly, it would depart from its orbit and slowly drift

A Bernal sphere design for a space colony. Space colonies may eventually be built, but it is unlikely they will be this sort of utopia. They will more likely contain technicians and construction workers in a boom town style economy and way of life. (Courtesy NASA)

away after a period of time. L_4 and L_5 lie at equal distances from the earth-moon system and form an equilateral triangle with the earth and moon. For purposes of definition, they are regarded as being in the moon's orbit.

It is possible to make all of the L-orbits stable, with some modifications. A space colony at L_5, for example, would be completely stable in relation to the earth, the moon, and the sun if it were in orbit around the Lagrangian point instead of exactly on it. Several space-colony designs place various components at different orbital points: a power satellite at one Lagrangian point, a receiver for lunar materials at another, and the actual colony at a third.

Some of the space colony plans approach gigantic sizes, with a potential for twenty thousand or more inhabitants. Some supporters of the space-colonies concept even think that it will be a future way of relieving population pressure on earth. Although the population pressure will increase for at least the next three generations, it is highly unlikely that space colonies can ever be a safety valve, because new births on earth would equal the colony's population long before it could be built in space. Nevertheless, a large space colony of technicians and scientists who would be the protectors of a future, giant space industry is not an impossible idea.

Whether these colonies would approach the idealistic dreams of vast, planned utopias, as envisioned by some supporters, is difficult to assess. Given man's history, it would appear very improbable. Western boom towns, for instance, although they contributed to the civilization and expansion of the West, would hardly qualify as utopian environments. Neither would the oil-boom construction towns and other societies on the front of a technological wave.

These concepts of space colonies were developed assuming only the lunar mining, space construction technology, and solar power satellites. One addition to the scenario is using the additional minerals in the asteroids. Meteorites which strike the earth often consist of very high-grade steel, and the asteroids are believed to be similarly composed. An iron-nickel asteroid a little over .5 mile in diameter (1 kilometer) would represent 4 billion tons of high-quality steel with a value of perhaps $500 billion.

Out in the asteroid belt, there are not only nickel-iron chunks of giant size but bodies rich in carbon and water, like the carbonaceous meteorites which strike earth. These, too, could be used as future supplies for space colonies, moon bases, and space factories.

Mining the asteroids is not necessarily a proposition for the far future. Using energy from the sun and fuel from an asteroid, small bodies could be moved from their original orbits to ones near the earth. There they would be mineral resources. as long as they last. Other and larger asteroids could be maneuvered into orbits near the earth to provide future mineral sources.

That asteroids could be used as raw materials is easily concluded from current mining operations on the earth's surface. A large area of Ontario, Canada, called the Sudbury Astrobleme, appears to have been formed by a prehistoric impact of a very large nickel-iron meteorite. Mining in the Ontario impact site yielded about one-half of the world's supply of nickel during the 1960s. The size difference between a large meteorite and a small asteroid is semantic only.

Of particular interest to those who advocate the mining of asteroids for raw materials is the group of asteroids called earth-grazers, bodies which approach very close to earth and are therefore easy to reach by current spacecraft technology. The earth-grazers come in two categories: Apollo asteroids which have orbits that intersect the earth's at some point, and Amor asteroids which lie between the orbit of Earth and the orbit of Mars. (The main asteroid belt is between Mars and Jupiter.) There are about forty known, close-approaching asteroids, varying from several hundred feet to 60 miles (100 kilometers) in diameter. They add up to plenty of raw material for a twenty-first century space economy and lunar colony, and enough for several space colonies.

During the development of the Space Transportation System, and the expansion into near-earth orbit and the geosynchronous orbits, the unmanned exploration of the solar system will have been continued. The major reason, of course, is scientific. But there are other reasons: the search for raw materials, mapping and investigating for future manned landings, sample-return missions to test soil potential and the existence of life forms in other places in the solar system.

Unmanned orbiter and lander spacecraft—in addition to robot hovering craft, that is, glider and lighter-than-air craft—will have visited all the planets and some of the major and minor asteroids. Missions will have been flown to the brightest comets, and small, long-duration spacecraft will have been sent on their way to the outer reaches of the sun's influence, to the edge of the solar wind in ultraplanetary space. Broadcasting from distances approaching tenths of a light year, these unmanned craft will investigate the area between the solar system's planets and the start of interstellar space between the sun and the nearest star.

By the time a lunar base is in operation, and space technology has progressed to large manned space stations, space factories, and construction techniques in high orbit, manned space exploration will have returned.

It has generally been assumed that the first manned interplanetary flight would be to Mars, and there is no reason to suppose that the situation has changed. Mercury represents a large expenditure of propulsion energy for every pound of payload delivered, and it is, after all, a singularly uninteresting planet, much like the moon. Mercury is, however, very rich in minerals and has vast scientific value. It is possible that the race for raw materials and minerals, which are even now in short supply on earth, would make Mercury more attractive as a manned mission candidate in the future.

Venus, Russian and American probes have shown, is about as inhospitable a planet as the solar system has. Recent investigations by an American Venus orbiter, the Pioneer Venus mission, indicate that it is even more inhospitable, if that is possible, than was formerly believed. Temperatures are on the order of 1000 degrees (540 Celsius), pressures are up to ninety times that on the surface of the earth, and the atmosphere has sulphuric acid clouds. The surface is extremely rugged and uninviting.

Mars has always fascinated mankind in one way or another, and since it was discovered that it was, in some ways, a sister planet of Earth with atmosphere,

A manned planetary vehicle might leave for Mars in this century; certainly by 2025 there will have been a manned planetary mission. (Courtesy NASA)

clouds, ice caps, and land masses, it has been the target of dreams. Life perhaps existed there, if it did nowhere else in the solar system. It had continents, what were once thought to be seas, and it had canals.

Even when unmanned spacecraft finally went to Mars and confirmed what had been long suspected—that there were no canals, that it was extremely cold, that the atmosphere had practically no oxygen, that there was little water, and apparently no life that could be detected—the dream of landing on the red dusty deserts did not die. It is still the most hospitable planet in the solar system, apart from Earth, and it is there that mankind will probably land, when a manned mission to another planet is designed and funded.

To go to Mars, funding is more significant than design. NASA designed a manned mission to Mars as early as 1969. Twelve astronauts could depart earth orbit November 12, 1981, and return to earth in August 1983. The original NASA Mars-mission plan called for two spaceships, which would reach Mars after going almost half-way around the sun in a giant circumsolar ellipse, taking 270 days for the trip. The expedition would arrive August 9, 1982, and enter an elliptical orbit around Mars.

After arriving in orbit, the two spaceships would remain there for eighty days, while six of the twelve crew members journeyed to the surface in landing vehicles. In October 1982, the ships would depart Mars orbit, cross the orbit of Earth and swing by the planet Venus to reduce their velocity for an Earth approach. The Venus swingby would take place at 123 days into the return mission and Earth approach would be at 167 days, on August 14, 1983.

For the original NASA Mars mission to have left on schedule, the initial go-ahead would have occurred in 1970 when the decision on the engine for the mission would have been made. A Mars excursion module would have been started in 1974. None of these decisions were ever made, so a manned Mars mission for 1981 is out of the question. There is no technology which would make it happen.

But for the future, with the development of new construction techniques and the possibility that a manned Mars vehicle could be built in near-earth orbit and return to a manned space station, a Mars mission is possible, at least in theory, every two years. Mars is in opposition or closest to Earth every two years. But some oppositions are much closer than others, because of the elliptical natures of the orbits of the two planets. A 1984 Mars mission has been suggested, as has one for 1988. Other possibilities are 1992 and 1996.

Beyond a manned Mars mission, there are other possibilities: the exploration of the asteroid belt, a mission to Venus, a long trek to one of the moons of the giant planet Jupiter. A mission may be possible to the curious moon of Saturn, Titan, which we know has an atmosphere, and which recent studies by unmanned space craft indicate may have frozen water on its surface in abundance.

Someday a manned base may be built on Mars in much the same way it could

In the far future, a manned Mars base has been established. Rovers and processors mine minerals for the bases. The Mars base would probably be nuclear powered. (Courtesy NASA)

be on the moon, beginning with a manned Mars orbiting station. And someday, there may come a time when the atmosphere of Mars will be made suitable for mankind through planetary engineering techniques called terraforming. Even inhospitable Venus may be changed by gigantic engineering programs into a planet where the race of mankind can go, explore, and pioneer.

The floor of the museum was slick and polished, but it bore the marks of a million feet deeply embedded in the rock surface. The air in the museum was clean, pure, but there was a faint odor of hot metal. It was late. A group of tourists stood in the center of the small gallery.

"Amazing," one said.

"I'd never have gone up in it," said another.

"They were heroes then," a man sighed, "not like us."

"I want to go home," whined a child.

As with all tourists there was not enough time. Vacation was short and getting shorter and they didn't want to spend forever on an afternoon in a musty museum surrounded by history.

The sun was still high in the sky, high over the museum; no one in the tourist group saw it. The ceiling was too thick and there were no windows. But the museum was closing on another day in just another ordinary week. In the smaller gallery, the visitors swam in the thin air and low gravity out to another exhibit and afterward to the airlock, hurried on by the impatient museum guides.

In the smaller gallery, colder now with the people gone, the gallery hollowed out of the rock of the moon, hung a silver, aluminum exhibit painstakingly pieced together, lovingly restored by the museum curators and technicians. The plaque below it read: SPACE SHUTTLE. 1980–1995. The World's First Spaceship. Lunar Space Museum.

Appendix A

NASA ABBREVIATIONS

AMPS	atmosphere, magnetosphere, and plasmas in space
APP	astrophysics payloads
ATL	advanced technology laboratory
CADSI	communications and data systems integration
CITE	cargo integration test equipment
DOD	Department of Defense
Domsat	domestic satellite
DSN	deep space network
ECS	environmental control system
ESA	European Space Agency
EVA	extravehicular activity
EVAL	earth-viewing application laboratory
GSFC	Goddard Space Flight Center
IPS	instrument pointing subsystem
IUS	interim upper stage
JPL	Jet Propulsion Laboratory
JSC	Lyndon B. Johnson Space Center
KSC	John F. Kennedy Space Center
LaRC	Langley Research Center

LDEF	Long Duration Exposure Facility
MCC	Mission Control Center (at JSC)
MMS	Multimission Modular Spacecraft
MSFC	Marshall Space Flight Center
NASCOM	NASA communications network
OMS	orbital maneuvering subsystem
OPF	Orbiter Processing Facility (at KSC)
OV	orbiter vehicle
PCR	payload changeout room (at launch site)
RCS	reaction control subsystem
RMS	remote manipulator system
SAEF-1	Spacecraft Assembly and Encapsulation Facility no. 1
SMS	shuttle mission simulator (at JSC)
SPP	solar physics payloads
SSUS	spinning solid upper stage
SSUS-A	spinning solid upper stage for Atlas-Centaur class spacecraft
SSUS-D	spinning solid upper stage for Delta class spacecraft
STDN	space tracking and data network
STS	Space Transportation System
TDRSS	Tracking and Data Relay Satellite system
VAB	Vehicle Assembly Building (at KSC)
VAFB	Vandenberg Air Force Base

Appendix B

SPACE SHUTTLE
DIAGRAMS

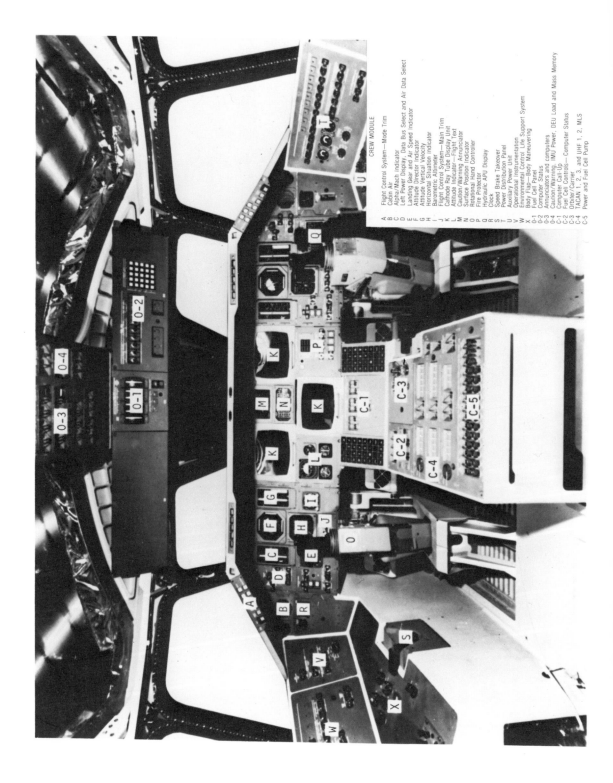

CREW MODULE

A Flight Control System—Mode Trim
B Cabin Air
C Alpha Mach Indicator
D Left Power Display, Data Bus Select and Air Data Select
E Landing Gear and Air Speed Indicator
F Attitude Director Indicator
G Attitude Vertical Velocity
H Horizontal Situation Indicator
I Barometric Altimeter
J Flight Control System—Main Trim
K Cathode Ray Tube Display Unit
L Attitude Indicator—Flight Text
M Caution Warning Annunciator
N Surface Position Indicator
O Rotational Hand Controller
P Fire Protector
Q Hydraulic APU Display
R Clock
S Speed Brake Takeover
T Power Distribution Panel
U Auxiliary Power Unit
V Operational Instrumentation
W Environmental Control Life Support System
X Body Flap—Body Maneuvering
O-1 Fuel Cell Panel
O-2 Computer Status
O-3 Annunciators and computers
O-4 Caution/Warning, IMU Power, DEU Load and Mass Memory
C-1 Computer Call-Up
C-2 Fuel Cell Controls—Computer Status
C-3 Orbiter/Carrier
C-4 TACAN 1, 2, 3, and UHF 1, 2, MLS
C-5 Power and Fuel Cell Pump

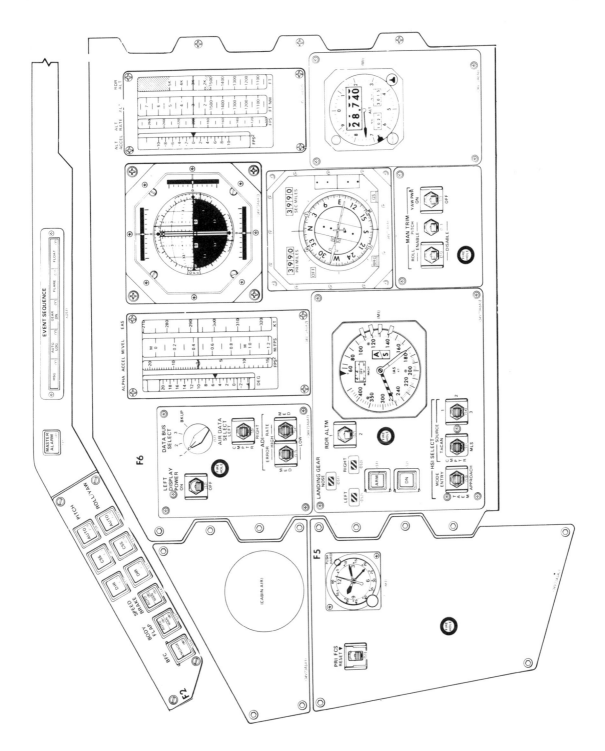

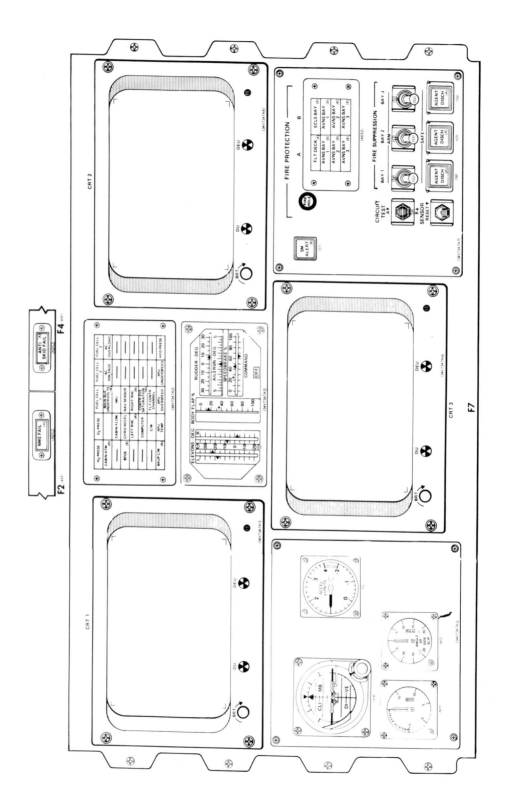

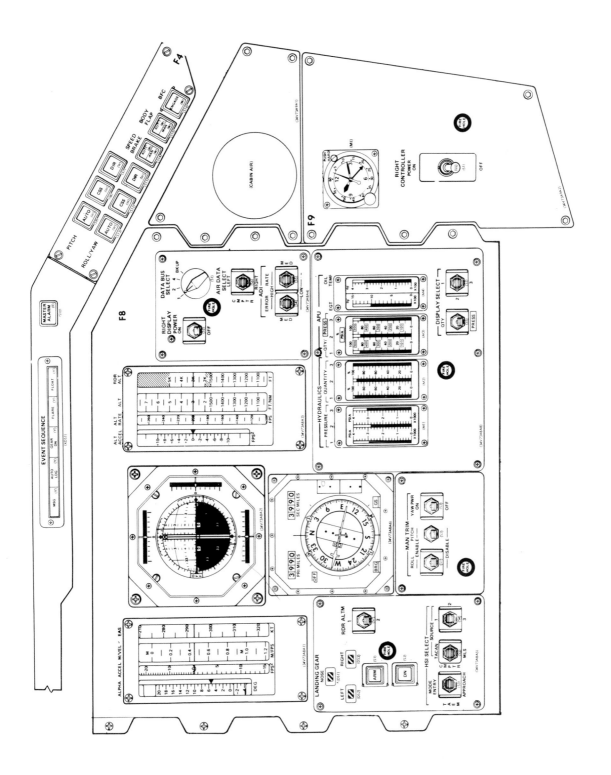

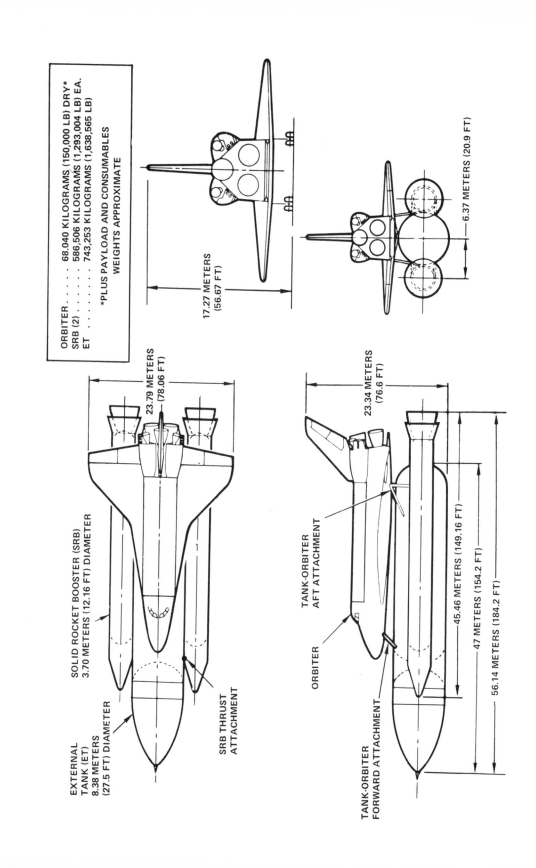

ORBITER . . . . . . 68.040 KILOGRAMS (150,000 LB) DRY*
SRB (2) . . . . 586,506 KILOGRAMS (1,293,004 LB) EA.
ET . . . . 743,253 KILOGRAMS (1,638,565 LB)

*PLUS PAYLOAD AND CONSUMABLES
WEIGHTS APPROXIMATE

17.27 METERS
(56.67 FT)

6.37 METERS (20.9 FT)

23.79 METERS
(78.06 FT)

EXTERNAL
TANK (ET)
8.38 METERS
(27.5 FT) DIAMETER

SOLID ROCKET BOOSTER (SRB)
3.70 METERS (12.16 FT) DIAMETER

SRB THRUST
ATTACHMENT

23.34 METERS
(76.6 FT)

TANK-ORBITER
AFT ATTACHMENT

ORBITER

TANK-ORBITER
FORWARD ATTACHMENT

45.46 METERS (149.16 FT)

47 METERS (154.2 FT)

56.14 METERS (184.2 FT)

SOLID ROCKET BOOSTER

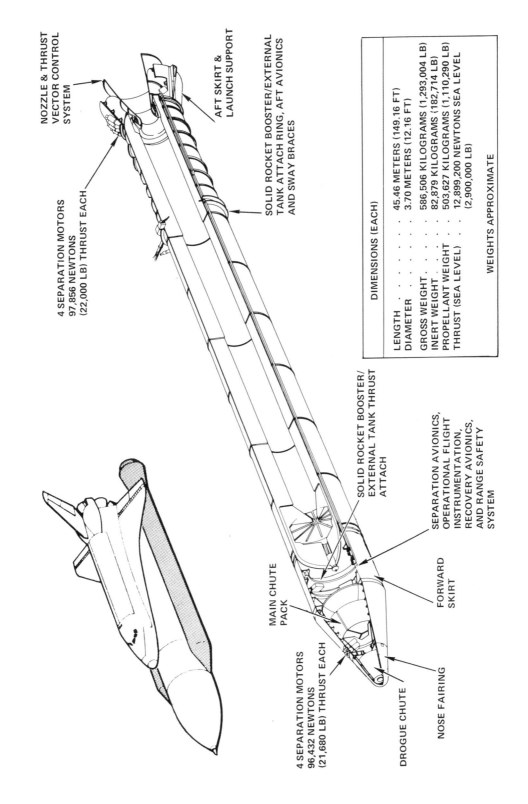

NOZZLE & THRUST VECTOR CONTROL SYSTEM

4 SEPARATION MOTORS 97,856 NEWTONS (22,000 LB) THRUST EACH

AFT SKIRT & LAUNCH SUPPORT

SOLID ROCKET BOOSTER/EXTERNAL TANK ATTACH RING, AFT AVIONICS AND SWAY BRACES

SOLID ROCKET BOOSTER/EXTERNAL TANK THRUST ATTACH

SEPARATION AVIONICS, OPERATIONAL FLIGHT INSTRUMENTATION, RECOVERY AVIONICS, AND RANGE SAFETY SYSTEM

FORWARD SKIRT

MAIN CHUTE PACK

4 SEPARATION MOTORS 96,432 NEWTONS (21,680 LB) THRUST EACH

DROGUE CHUTE

NOSE FAIRING

DIMENSIONS (EACH)	
LENGTH	45.46 METERS (149.16 FT)
DIAMETER	3.70 METERS (12.16 FT)
GROSS WEIGHT	586,506 KILOGRAMS (1,293,004 LB)
INERT WEIGHT	82,879 KILOGRAMS (182,714 LB)
PROPELLANT WEIGHT	503,627 KILOGRAMS (1,110,290 LB)
THRUST (SEA LEVEL)	12,899,200 NEWTONS SEA LEVEL (2,900,000 LB)

WEIGHTS APPROXIMATE

EXTERNAL TANK

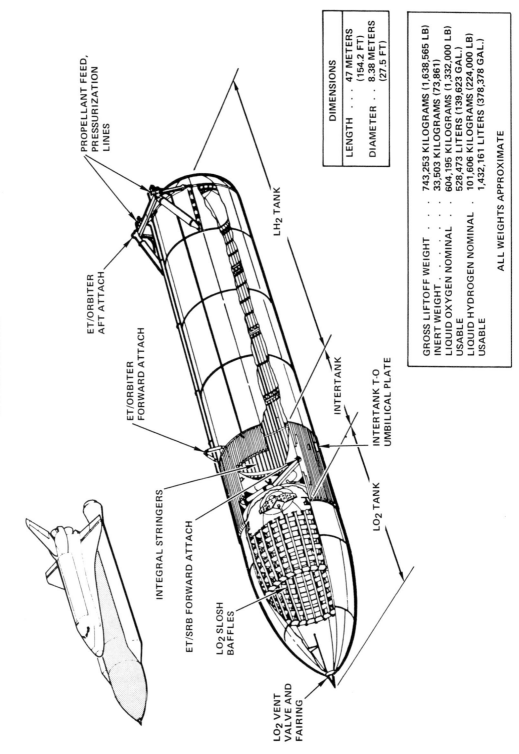

PROPELLANT FEED, PRESSURIZATION LINES

ET/ORBITER AFT ATTACH

ET/ORBITER FORWARD ATTACH

LH₂ TANK

INTEGRAL STRINGERS

ET/SRB FORWARD ATTACH

INTERTANK

INTERTANK T-O UMBILICAL PLATE

LO₂ TANK

LO₂ SLOSH BAFFLES

LO₂ VENT VALVE AND FAIRING

DIMENSIONS	
LENGTH . . .	47 METERS (154.2 FT)
DIAMETER . .	8.38 METERS (27.5 FT)

GROSS LIFTOFF WEIGHT	743,253 KILOGRAMS (1,638,565 LB)
INERT WEIGHT	33,503 KILOGRAMS (73,861)
LIQUID OXYGEN NOMINAL . .	604,195 KILOGRAMS (1,332,000 LB)
USABLE	528,473 LITERS (139,623 GAL.)
LIQUID HYDROGEN NOMINAL .	101,606 KILOGRAMS (224,000 LB)
USABLE	1,432,161 LITERS (378,378 GAL.)

ALL WEIGHTS APPROXIMATE

THERMAL PROTECTION SYSTEM

Legend:
- REINFORCED CARBON-CARBON (RCC)
- HIGH-TEMPERATURE, REUSABLE (HRSI) SURFACE INSULATION
- LOW-TEMPERATURE, REUSABLE (LRSI) SURFACE INSULATION
- COATED NOMEX FELT (FRSI) REUSABLE SURFACE INSULATION
- METAL OR GLASS

ORBITER 102 CONFIGURATION

TPS (THERMAL PROTECTION SYSTEM)*	AREA		WEIGHT	
	SQUARE FEET	SQUARE METERS	POUNDS	KILOGRAMS
FRSI	3436	319	1099	499
LRSI	2857	265	2256	1023
HRSI	5172	481	9666	4385
RCC	409	38	2913	1321
MISCELLANEOUS			1398	634
TOTAL	11874	1103	17332	7861

*INCLUDES BULK INSULATION, THERMAL BARRIERS, AND CLOSEOUTS

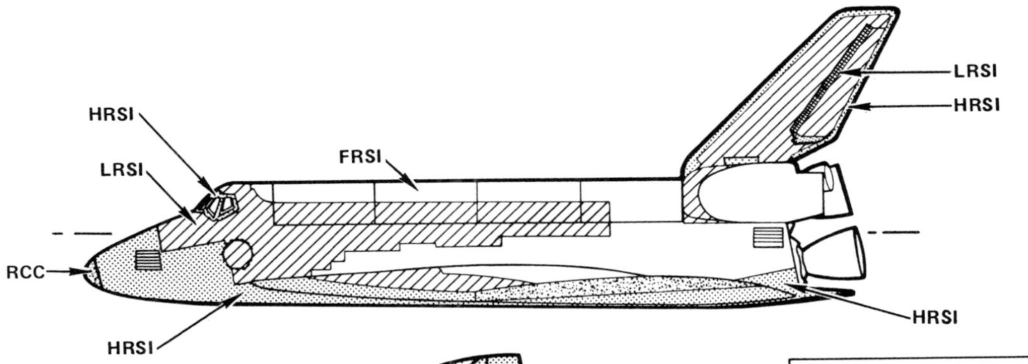

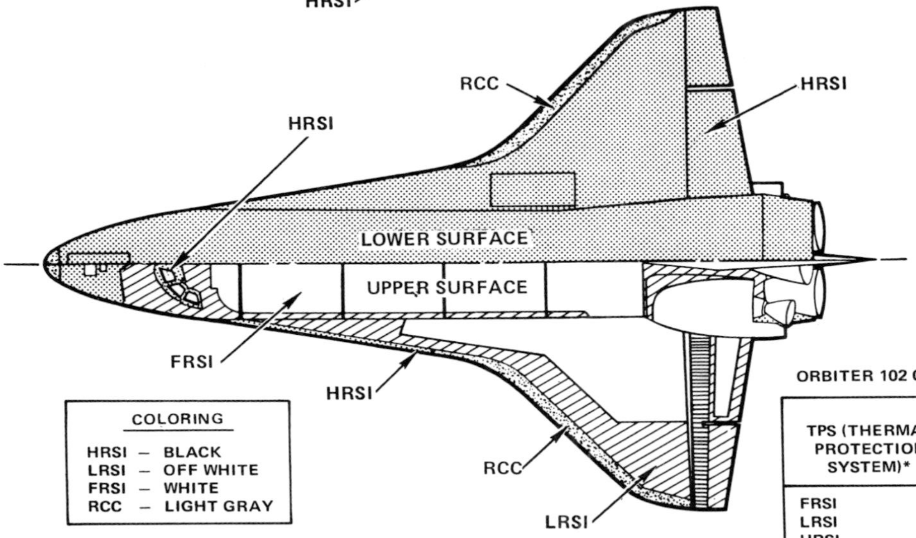

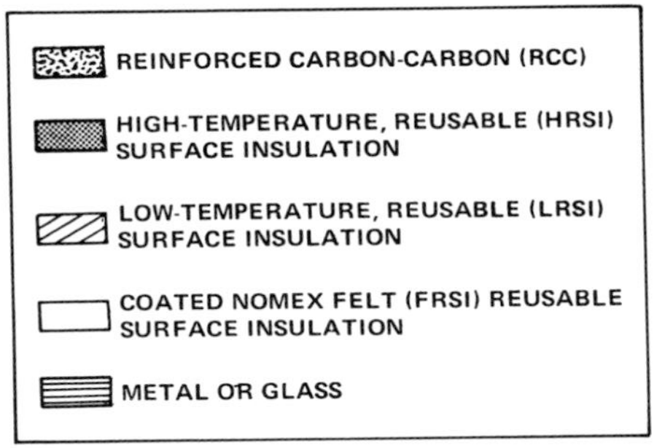

COLORING

HRSI — BLACK
LRSI — OFF WHITE
FRSI — WHITE
RCC — LIGHT GRAY

LRSI HRSI HRSI FRSI HRSI LRSI RCC HRSI LOWER SURFACE UPPER SURFACE HRSI LRSI RCC HRSI FRSI RCC

SPACE SHUTTLE MAIN ENGINES

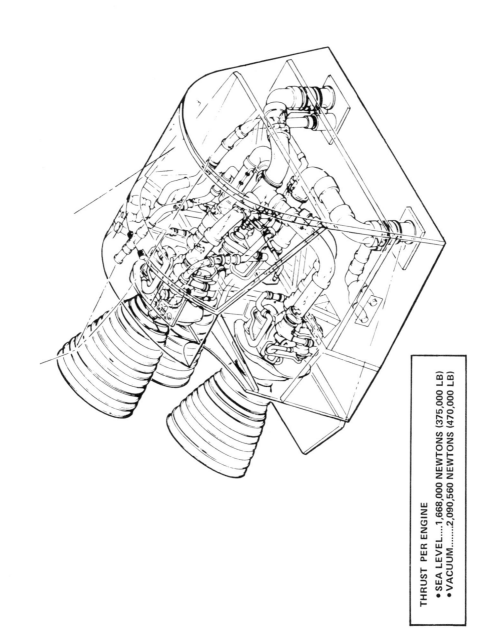

THRUST PER ENGINE
- SEA LEVEL.....1,668,000 NEWTONS (375,000 LB)
- VACUUM.......2,090,560 NEWTONS (470,000 LB)

REACTION CONTROL SYSTEM

1 FORWARD RCS MODULE, 2 AFT RCS PODS
38 PRIMARY THRUSTERS (14 FORWARD, 12 PER AFT POD)
  THRUST LEVEL = 3,870 NEWTONS (870 LB) VACUUM
6 VERNIER THRUSTERS (2 FORWARD, 4 AFT)
  THRUST LEVEL = 111.2 NEWTONS (25 LB) VACUUM
PROPELLANTS: NITROGEN TETROXIDE - OXIDIZER
              MONOMETHYL HYDRAZINE - FUEL

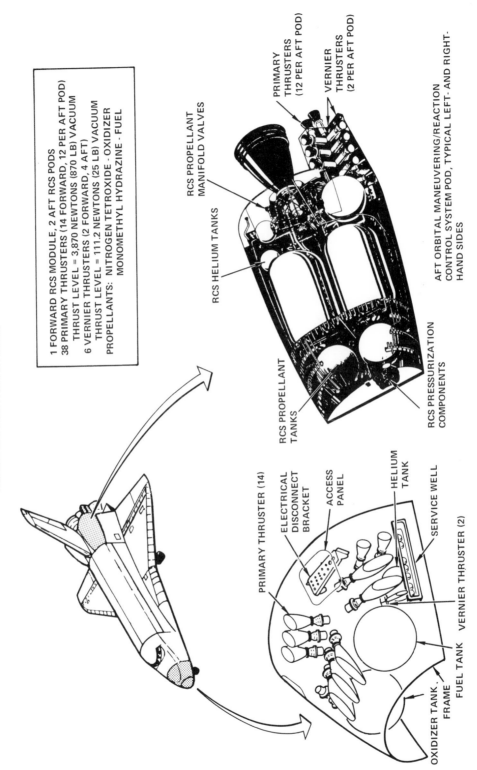

RCS PROPELLANT
MANIFOLD VALVES

RCS HELIUM TANKS

RCS PROPELLANT
TANKS

RCS PRESSURIZATION
COMPONENTS

PRIMARY
THRUSTERS
(12 PER AFT POD)

VERNIER
THRUSTERS
(2 PER AFT POD)

AFT ORBITAL MANEUVERING/REACTION
CONTROL SYSTEM POD, TYPICAL LEFT- AND RIGHT-
HAND SIDES

PRIMARY THRUSTER (14)

ELECTRICAL
DISCONNECT
BRACKET

ACCESS
PANEL

HELIUM
TANK

SERVICE WELL

VERNIER THRUSTER (2)

OXIDIZER TANK
FRAME

FUEL TANK

FORWARD REACTION CONTROL SYSTEM

ORBITAL MANEUVERING SYSTEM

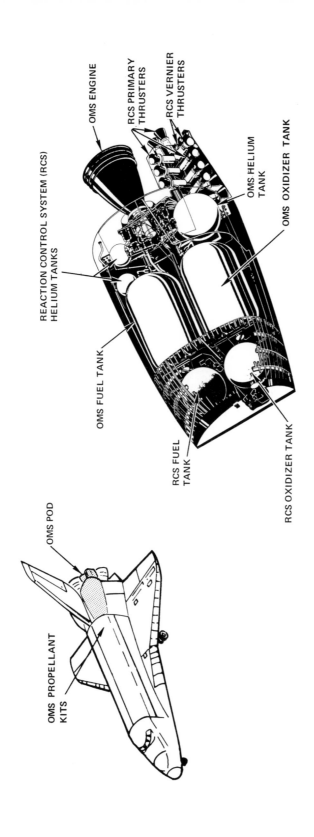

OMS ENGINE

OMS ENGINE	
THRUST 26,688 NEWTONS (6,000 LB) VACUUM	
PROPELLANT (PER POD)	
MONOMETHYL HYDRAZINE FUEL 2,043 KILOGRAMS (4,505 LB) & NITROGEN TETROXIDE OXIDIZER 3,372 KILOGRAMS (7,433 LB)	} USABLE
WEIGHTS APPROXIMATE	

GLOSSARY

aft flight deck: part of the upper deck of the orbiter cabin, where the payload operating controls are located.

airlock: a passage between a pressurized and an unpressurized portion of a spacecraft, or between a pressurized compartment in a spacecraft and the vacuum of space outside. Crew members in the orbiter move through an airlock to exit into space for EVAs. The spacelab module uses an airlock to expose experiments to outer space.

Ames Research Center: a NASA establishment at Moffett Field, California. It was set up in 1940 by the predecessor of NASA, the National Advisory Committee for Aeronautics (NACA).

Anik: the Anik satellites, also called Telesat, are communications satellites for Canada. They are in synchronous orbits and therefore are stationary over 104 degrees, 109 degrees, and 114 degrees west longitude. Each satellite carries twelve color TV channels and 5760 telephone circuits.

Atlas-Centaur class: a payload class weighing 4000 to 4400 pounds (1800 to 2000 kilograms).

Apollo-Soyuz Test Project: also known as ASTP. It was a joint Soviet-American mission, involving an *Apollo* spacecraft docking with a Russian Soyuz spacecraft. Coming partly from this ASTP mission was the design for the international docking module for the space shuttle.

asteroid: a small, rocky body moving around the sun in a highly elliptical orbit. They are also called planetoids and minor planets. Thousands are known to exist with diameters ranging from 600 miles (1000 kilometers) down to rocks the size of houses and below. Debris from asteroids sometimes hits the earth and is called a meteorite.

astrophysics: the application of physics to the study of the universe.

attitude: the orientation of a spacecraft (or an aircraft) to a given frame of reference, for example, the local direction of the vertical.

automatic landing mode: the space shuttle has a computer-controlled guidance and control system, making it capable of landing completely automatically.

azimuth: true launch heading from Kennedy Space Center, or Vandenberg Air Force Base, measured clockwise from zero degrees north.

barbecue mode: the space shuttle orbiter doing a slow roll to counteract external heat.

beta cloth: the material from which spacesuits are made. It is flameproof and made of glass fibers.

booster: sometimes used to describe the first stage of a rocket. Strap-on boosters are used by Titan III rockets and by the space shuttle, where they are called SRBs.

bulkhead: a wall.

Canopus: a bright, southern constellation star. It is most often used to orient spacecraft, including the orbiter. It is easily located because of its brightness.

capture: a payload is captured by the shuttle orbiter anytime the payload is firmly attached to the remote manipulator arm.

cargo: on the space shuttle, cargo is the total complement of payloads on any one flight. It includes everything in the orbiter cargo bay, plus other equipment, hardware, and consumables located elsewhere in the orbiter.

cargo bay: the unpressurized middle part of the space shuttle orbiter fuselage. It has hinged doors which extend the full length of the cargo bay.

commander: the person ultimately responsible for the safety of space shuttle personnel, who has authority throughout the flight to deviate from the flight plan and procedures.

core segment: a section of the pressurized spacelab module which houses subsystem equipment and experiments.

crew egress: NASA-talk for leaving the orbiter after landing.

crew ingress: NASA-talk for getting on board the orbiter.

deck: the floor levels of the orbiter: flight deck, mid-deck, and lower deck. There is no up or down in space, but the deck is considered as being down for reference.

delta wing: a triangular wing. The orbiter is regarded as a delta-winged aircraft (spacecraft).

deep space network: Jet Propulsion Laboratory's communications network for command and control of all planetary flights.

Delta class: payloads weighing 2000 to 2500 pounds (900 to 1100 kilograms).

deorbit burn: a rocket thrust period which reduces the space shuttle's orbital speed. It can change orbits or begin the reentry and landing sequence.

deployment: the process of removing a payload from a berthed position in the shuttle cargo bay and releasing the payload into space.

entry: the time period when a spacecraft first encounters the upper atmosphere of the earth. Also called reentry.

Earth Resources Technology Satellite: the original name for NASA's Landsat satellite.

European Space Agency: an international organization of eleven nations: Belgium, Denmark, Eire (Ireland), France, West Germany, Italy, Netherlands, Spain, Sweden, Switzerland, and the United Kingdom. It was organized for space research, rocket development, and satellite applications. ESA is responsible for the construction of the spacelab payload of the Space Transportation System.

external tank: the aluminum fuel tank for the orbiter. It is 154 feet (46.9 kilometers) long and 28 feet (8.5 meters) in diameter and stores liquid oxygen and liquid hydrogen for use by the orbiter's main engines.

extravehicular activity: activities by shuttle crew outside the pressure hull, or within the cargo bay, when the cargo bay doors are open.

extravehicular mobility unit: a self-contained, life support system and pressure suit for use during extravehicular activity.

flight: the period from launch to landing of an orbiter. A single space shuttle round trip. Not the same as a mission; a shuttle mission may require more than one flight.

flame trench: a concrete pit under a launch pad. It directs the fierce rocket exhaust away from the shuttle and launch facilities during liftoff.

free-flying system: any satellite or payload which is detached from the orbiter and is capable of independent operation.

g: a unit of force exerted on a body by gravity. The earth's gravity exerts a force of one g. A three-g acceleration force is three times that exerted by earth's gravity. An astronaut experiencing three-g acceleration would feel as if he or she weighed three times more than normally.

gamma ray astronomy: study of wavelengths with shorter radiation than X rays.

glide slope: the final landing approach of the space shuttle during its landing phase.

hypergolic propellants: propellants which ignite spontaneously on contact with each other. They do not require an igniter. Nitrogen tetroxide and monomethylhydrazine are used in space shuttle.

Intelsat: International Telecommunications Satellite Corporation. An organization of a number of nations, formed in 1964, to produce and operate international communications satellites.

igloo: a pressurized container for the spacelab pallet subsystems when no manned, pressurized module is used for a flight.

integration: assembling payloads and space transportation components into a desired configuration.

interim upper stage: a solid-propellant upper stage designed to put spacecraft into high-earth orbits, or on escape trajectories for planetary missions.

Jet Propulsion Laboratory: a NASA division in Pasadena, California. It was set up in 1944 for missile development and joined NASA in 1958.

launch pad: area at which space shuttle, or any other rocket or spacecraft, is launched from the surface of the earth.

long duration exposure facility: a free-flying, reuseable satellite, designed primarily for small passive, or self-contained active, experiments that require long exposure to space. It is launched in the orbiter cargo bay, deployed and retrieved by the remote manipulator arm. The LDEF may be left in space and retrieved during a later shuttle flight.

LOX: short for liquid oxygen.

manned maneuvering unit: a propulsive backpack for extravehicular activity.

mass: a basic property of matter. When mass is accelerated by gravity or by the thrust of a rocket, there is the sensation of weight.

mission: a shuttle program with specific goals. A mission might require more than one shuttle flight.

Mission Control: the mission control center at Johnson Space Center for control and support of all phases of space shuttle flights.

mission-dependent equipment: spacelab optional equipment which can be added to a flight when necessary.

mission-independent equipment: spacelab subsystem and support equipment which is carried on every flight.

mission specialist: a space shuttle crew member who is responsible for the coordination of overall payload/space shuttle interaction. During the payload-operations phase, the mission specialist directs the space shuttle and crew resources to accomplish the required payload objectives.

mission station: a control station on the orbiter aft flight deck from which payload support operations are performed.

mobile launch platform: the structure on which the elements of the shuttle are stacked in the Vehicle Assembly Building and are moved to the launch pad.

module: a pressurized, manned laboratory suitable for science applications and technological activities. It also refers to large or small plug-in sections for space structures or a space system.

multimission modular spacecraft: a free-flying system built in sections so that it can be adapted to many missions requiring earth-orbiting, remote-sensing spacecraft. It is launched from the orbiter cargo bay and deployed and retrieved by the shuttle's remote manipulator arm.

orbital flight test: one of the six scheduled developmental space flights of the space shuttle.

orbital maneuvering system: orbiter engines which provide the thrust to perform orbit insertion, circularization, or transfer; rendezvous; and deorbiting.

orbiter processing facility: a building near the Vehicle Assembly Building at the Kennedy Space Center, with two bays in which the orbiter is placed for inspection, maintenance, and checkout prior to the installation of a payload. Payloads are horizontally installed in this building.

pallet: an unpressurized platform, designed for installation in the orbiter cargo bay.

Instruments and equipment requiring direct space exposure will be mounted on pallets.

pallet train: more than one pallet, rigidly connected to form a single unit.

payload changeout room: an environmentally controlled room at the launch pad for inserting payloads vertically into the orbiter cargo bay.

payload specialist: a shuttle crew member who may or may not be a career astronaut. The payload specialist is responsible for the operation and management of specifically assigned experiments or other payload elements.

payload station: a location on the aft flight deck of the orbiter, from which specific payload operations are carried out. The operations are done by a mission or payload specialist.

pilot: a crew member who is second-in-command of an orbiter flight.

reaction control subsystem: thrusters on the shuttle orbiter used to provide attitude control during orbit insertion, on-orbit, and reentry phases of a shuttle flight.

refurbishment: repair maintenance, replacement of parts, and general rebuilding of any device.

remote manipulator system: a mechanical arm on the cargo bay, controlled from the orbiter aft flight deck. It is used to deploy, retrieve, or move payloads.

solid rocket boosters: two recoverable, solid-propellant booster rockets used to help boost the shuttle off the launch pad and up to an altitude of about twenty-five miles.

Space Transportation System: the space shuttle (orbiter, external tank, solid rocket booster), upper stages, spacelab, and any associated flight hardware or software.

spinning solid upper stage: an upper stage designed to deliver Delta and Atlas Centaur spacecraft classes to earth orbits beyond the capabilities of the space shuttle.

tilt/spin table: a mechanism installed in the orbiter cargo bay which deploys the spinning solid upper stage with its attached spacecraft.

Tracking and Data Relay Satellite System: a two-satellite communications system which provides the principal coverage from geosynchronous orbit for all space shuttle flights.

upper stage: a spinning solid upper stage or an interim upper stage. Both are designed for launch from the orbiter cargo bay to deliver payloads into orbits beyond the capabilities of the shuttle orbiter.

Vehicle Assembly Building: a high-bay building near the Kennedy Space Center launch pad in which shuttle elements are stacked onto the mobile launch platform. It is also used for vertical storage of the external tanks.

Western Launch Operations: NASA operations at Vandenberg Air Force Base, California.

ADDITIONAL READINGS

Asimov, Isaac. "After Apollo, A Colony on the Moon." New York *Times Magazine* 30, May 28, 1967.

———"Colonizing the Heavens." *Saturday Review* 17, June 28, 1975.

———*Mars, The Red Planet*. Lothrop, Lee & Shepard Company, 1977.

———"The Next Frontier?" *National Geographic Magazine*, July 1976, pp. 76–89.

Bekey, Ivan, and Mayer, H. L. "1980–2000: Raising Our Sights for Advanced Space Systems." *Astronautics and Aeronautics,* July/August 1976.

Berry, Adrian. *The Next Ten-Thousand Years: A Vision of Man's Future in the Universe.* Saturday Review Press, 1974.

Blagonravov, Anatolij A. "Space Platforms: Why They Should be Built." *Space World* H-8-92, August 1971.

Bono, P., and Gatland, K. W. "The Commercial Space Station." *Frontiers of Space.* MacMillan Co., 1970.

Brand, Stewart, ed. *Space Colonies*. Penguin, 1977.

Bredt, J. H., and Montgomery, B. O. "Materials Processing in Space—New Challenges for Industry. *Astronautics and Aeronautics* 13, May 1975.

Chapman, Clark R. *The Inner Planets*. Charles Scribner's Sons, 1977.

Chedd, Graham. "Colonization at Lagrangia." *New Scientist,* October 24, 1974.

Chernow, Ron. "Colonies in Space." *Smithsonian,* February 1976, pp. 6–11.

Clarke, Arthur C. *Islands in the Sky*. NAL Signet Book, 1965.

Cole, Dandridge M. *Beyond Tomorrow: The Next 50 Years in Space*. Amherst Press, 1965.

Collins, Michael. *Carrying the Fire*. Farrar, Straus, and Giroux, 1974.

Cravens, Gwyneth. "The Garden of Feasibility." *Harper's,* August 1975.

Dempewolf, Richard. "Cities in the Sky." *Popular Mechanics,* 205, May 1975, pp. 94–97.

Dossey, J. R., and Trotti, G. L. "Counterpoint—A Lunar Colony." *Spaceflight* 17, July 1975.

Firsoff, V. A., *The Solar Planets*. Crane, Russak & Company, Inc., 1977.

Friedman, Phillip. "Colonies in Space." *New Engineer,* November 1975.

Glaser, Peter E. "Power from the Sun: Its Future." *Science* 162, November 22, 1968, pp. 857–861.

Guillen, Michael. "Moon Mines, Space Factories and Colony L-5." *Science News*, August 21, 1976.

Glenn, Jerome Clayton, and Robinson, George S. *Space Trek: The Endless Migration*. Stackpole Books, 1978.

Hagler, Thomas A. "Building Large Structures in Space." *Astronautics and Aeronautics* 14, May 1976, pp. 56–61.

Heppenheimer, Thomas A. *Colonies in Space*. Stackpole Books, 1977.

Kahn, German et al. *The Next Two Hundred Years*. Morrow, 1976.

Lessing, L. "Why the Shuttle Makes Sense." *Fortune* 85, January 1972, p. 93.

Lewis, Richard S. *The Voyages of Apollo*. Quadrangle, The N. Y. Times Book Co., 1974.

Ley, Willy. *Rockets, Missiles, and Space Travel*. The Viking Press, 1957.

Libassi, Paul T. "Space to Grow." *The Sciences,* July 8, 1974.

Michaud, M.A.G. "Escaping the Limits to Growth." *Spaceflight,* April 1975.

Moore, Patrick. *The Next Fifty Years in Space*. William Luscombe, 1976.

O'Leary, Brian T. "Asteorid Mining," *Astronomy,* November 1978.

O'Neill, Gerard K. "Colonies in Orbit." *New York Times Magazine,* January 18, 1976, p. 10.

————"The Colonization of Space." *Physics Today* 27, September 1974, pp. 32–40.

————*The High Frontier: Human Colonies in Space*. William Morrow and Co., 1977.

————"Space Colonies and Energy Supply to the Earth." *Science* 190, December 5, 1975, pp. 943–947.

Ordway, Frederick I. "The Advance of Rocket Science," *Sky and Telescope,* October 1952.

"Our Next 25 Years in Space," *Sky and Telescope,* November 1975.

Paine, Thomas. "Colonies in Space." *Time Magazine,* June 3, 1974.

Parkinson, R. C. "Takeoff Point for a Lunar Colony." *Spaceflight,* September 1974.

Powers, Robert M. *Planetary Encounters*. Stackpole Books, 1978.

Reis, Richard. "Colonization of Space." *Mercury* 3–4, July 8, 1974.

Robinson, George S. *Living in Outer Space*. Public Affairs Press, 1975.

Ridpath, Ian, ed. *The Encyclopedia of Astronomy and Space,* Thomas Y. Crowell Company, 1976.

Rosendhal, Jeffrey D. "The Spacelab 2 Mission." *Sky and Telescope,* June 1978.

Roth, Günter D. *The System of Minor Planets*. Van Nostrand, 1962.

Stine, G. Harry. *The Third Industrial Revolution*. G. P. Putnam's Sons, 1975.

Vondrak, R. "Creation of an Artificial Lunar Atmosphere." *Nature,* March 1974.

Williams, J. R. "Geosynchronous Satellite Solar Power." *Astronautics and Aeronautics* 13, November, 1975, pp. 46–52.

INDEX

Advanced Programs Division, Office of Space Transportation Systems (NASA), 173
Air Force Space Museum, Cape Canaveral, 86
Airlock, shuttle orbiter, 73; airlock tunnel, spacelab, 150 (ill.)
Aldrin, E. E., Jr., 19
Antennas: communications, on shuttle orbiter, 84; for electronic mail, 193, 195 (ill.); for extraterrestrial intelligence, 191 (ill.), 194; for personal communication, 193
Apollo, 19, 66, 75, 106–7, 122, 187; *Apollo 8,* 80–81; *Apollo 17,* 110; *Apollo-Soyuz,* 19, 81, 110, 187; *Saturn,* 89; *Saturn V,* 25, 39, 44, 96
Armstrong, Neil, 19
Asteroids, 224–25; mining of, 224
Astronomy satellite, small (SAS-2), 166
Atlas, 44; Atlas-Agena, 44, 86; Atlas-Centaur, 86, 137–38
Atmosphere, magnetosphere, and plasmas in space (AMPS), spacelab, 144 (ill.), 151, 155 (ill.)

Beam builder, automated (ABB), 205 (ill.), 208; capacity of, 208
Bean, Alan, 109
Bernal sphere, for space colony, 223 (ill.)
Boeing Company, 45, 207
Bonestell, Chesley, 43
Boosters, solid rocket (SRBs); *see* Solid rocket boosters
Braun, Wernher von, 39–44, 46, 61–62, 85, 107, 158, 172
Brown, Edmund G., 181

Cabin, shuttle orbiter, 69, 78 (ill.)
Cape Canaveral, 20–21, 33, 43, 48, 51, 54, 56, 62, 86, 89, 125, 138; Cape Approach, 34; Air Force Station, 87, 89, 101
Cargo-payload bay, shuttle orbiter, 83–84
Clarke, Arthur C., 85, 107, 135
Clarke orbit, 135, 177
Close Encounters of the Third Kind, 121
Collins, Michael, 19

Colonies, space; *see* Space colonies
Communication Satellite Corp. (COMSAT), 124–25
Communications satellites (Comsats), 182, Communications Technology Satellite, 122, 124; personal communications, 193; Signal Communication by Orbiting Relay Equipment (SCORE), 182; *Telstar,* 86, 182; *Telstar Canada,* 101
Construction base, concepts of, 173, 179
Cooper, Gordon, 104
Copernicus satellite (OAO-3), 158
Countdown, origin of, 23
Crawler-transports, 96, 100 (ill.)
Crews: on shuttle orbiter, 62, duties of, 113, number of, 106, selection of, 106, 109–10, 113–14, survival training for, 109, training of, 115–16, 118, women in, 106, 108–10, 116; on *Spacelab 1,* 114
Cronkite, Walter, 107

Deorbit, of orbital maneuvering system, 28–30

Descent, 28–30; altitude of, 33–34; speed of, 33
Disher, John H., 197
Docking, shuttle, 189 (ill.)
Docking and rendezvous station, on shuttle orbiter, 73–74
Dryden Flight Research Center, 33
Dumas, Alexandre, 36

Edwards Air Force Base, 46, 62, 89
Ehricke, Kraft, 179
Electrical power, on shuttle orbiter, 83; fuel for, 83
Engines; *see* Space shuttle main engines
Engle, Joseph, 67
Enterprise; see Shuttle
Environmental control system, on shuttle orbiter, 73
European Space Agency, 144–45
Exercise, on shuttle orbiter, 77
External tanks (ET), 49 (ill.), 54, 178 (ill.); forward section, 55 (ill.); separation from shuttle, 54; size of, 51; structure of, 52–53 (ill.)
Extravehicular mobility unit, 71 (ill.), 73

Flightdeck, shuttle orbiter, 68, 68 (ill.), 73; organization of, 73
Food, on shuttle orbiter, 79–81, 83; food tray, 82 (ill.); storage container for, 80 (ill.)
Fuel loading, 20
Fuel tank; separation from orbiter, 25–26

Galileo orbiter probe, 128, 132, 136, 167 (ill.), 169, 192; launch date, 169
Galley, on shuttle orbiter, 82–83
Gamma Ray Observatory (GRO), 166
Gamma-ray satellite, 165
Garriott, Dr. Owen K., 109, 114
General Dynamics, 45
Gibson, Edward, 109
Glaser, Dr. Peter, 198
Global Positioning System program, 184
Goddard, Dr. Robert Hutchings, 38–39, 43, 46
Godwin, Bishop Francis, 36
Grumman Aerospace Corp., 45

Haise, Fred W., Jr., 111
Haldane, John B. S., 157
Hale, Edward Everett, 36
Hamilton Standard, 116
Hatch, on shuttle orbiter, 69
Heat shield, 30
Heinlein, Robert A., 171
Heppenheimer, T. A., 222
High-Energy Astronomical Observatory (HEAO), launches, 166
Housekeeping, on shuttle orbiter, 77
Hygiene, on shuttle orbiter, 79

Industrialization of space, 181–96; commercialization of, 188; electronic mail, 193, 195 (ill.); factories, 188, 189 (ill.); medicine, 188; personal communications antennas, 193; search for extraterrestrial intelligence, 191–92, 194; solar array wing, 186 (ill.); structures, types needed, 192; tourism, 188
Intelsat satellite, 182
Interim upper stage (IUS), 130–33, 131 (ill.), 133 (ill.), 135
International Sun-Earth Explorer, 124
International Ultraviolet Explorer, 125

Lyndon B. Johnson Space Center, 25, 34, 103, 111, 113
Jupiter, 86; Jupiter-C rocket, 44; Jupiter *Galileo* orbiter probe, 128, 132, 136

John F. Kennedy Space Center, 19–20, 30–31, 34, 86, 89–90, 101, 147
Kepler, Johann, 36
Kerwin, Dr. Joseph, 109

Lander, lunar, 212, 217
Lagrangian points, 222–23
Landing, 31–34, 59; Orbiter Landing Facility, 95; postlanding procedures, 90, 94–96; speed of, 32
Landing sites, Kennedy Space Center, 88 (ill.), 96
Landing strip, Kennedy Space Center, 93 (ill.)
Landsat satellites, 183 (ill.), 184

Lang, Fritz, 23
Langley Research Center, 127
Lasswitz, Kurd, 36
Launch, 23; preparation for, 20; shuttle on launch pad, 99 (ill.)
Launch pads, Cape Canaveral, 86, 89, 96; shuttle on, 99 (ill.)
Launch sites, Cape Canaveral, 86, 89
Ley, Willy, 42
Liftoff, 21–23
Arthur D. Little, Inc., 198
Long Duration Exposure Facility, 101, 125, 126 (ill.), 127
Lunar base, manned, 212, 217, 220, 222; cost of, 220; estimated date of first, 220
Lunar lander, 212, 217
Lunar mining, 222, 224
Lunar module stations, 217
Lunar orbiter, 212
Lunar polar orbiter, 136
Lunar rover, 216, 218 (ill.), 221 (ill.)
Lunar stations, manned, 212, 214–15 (ill.)
Lunar water-mining and processing station, 216 (ill.), 217

Maneuvering unit, manned, shuttle orbiter, 72 (ill.)
Manipulator arm, 84, 131 (ill.)
Manned orbiting laboratory (Air Force), 107
Mars, 225, 228; mission to, 225; theoretical manned base on, 229, 229 (ill.)
Marshall Space Flight Center, 108, 115
Martin Marietta, 45
materials science laboratory, spacelab, 153
McDonnell-Douglas Corp., 45
Mead, Margaret, 141
Meals, on shuttle orbiter, 79–83; food tray, 82 (ill.); storage container for, 80 (ill.)
Microwave conversion, solar-power satellite, 198, 203
Microwave receiving station, 203, 203 (ill.)
Mining, asteroids, 224; lunar, 222, 224
Mission station, on shuttle orbiter, 74
Modules: crew, shuttle orbiter, 66 (ill.); habitability, 177 (ill.); long, 148–49; power, shuttle orbiter, 138, 139 (ill.), 140, 176 (ill.)
Module stations, lunar, 217
Moon surface vehicle; *see* Rover

Motors, separation, on shuttle, 50
Multimission modular spacecraft, 127–28

National Aeronautics and Space Administration (NASA), 44, 48, 62, 65, 79, 86–87, 103, 106, 109, 112–16, 120, 122, 124–25, 127, 130, 132–33, 136–37, 144–45, 160, 169, 172–72, 199, 207, 212, 217, 228
National Space Technology Laboratory (NASA), 56
Navigation satellites, 184
NavStars 1, 2, and *3,* 184
Nimbus weather satellite, 184
Nimoy, Leonard, 19
Noordung, Hermann, 42
North American Rockwell Corp., 45, 212

Oberth, Hermann Julius, 39, 105, 158
O'Neill, Dr. G. K., 222
Orbital astronomical observatory (OAO), 158
Orbital Maneuvering System (OMS), 27; firing of, 28; deorbit of, 28–30; engines, 56; fuel for, 25; fuel propellant, 56; ignition, for deorbiting, 30; purpose of, 56; speed of, 30
Orbital transfer vehicle (OTV), 213 (ill.)
Orbital Aeroflight Simulator, 116
Orbiter, lunar, 212
Orbiter Processing Facility, 93 (ill.), 95 (ill.), 96
Orbiter, shuttle, 49 (ill.), 58 (ill.), 64–65 (ill.); airlock, 73; cabin of, 69, 78; cargo-payload bay, 83–84; crew, number of, 62; crew module, 66 (ill.); communications antennas, 84; components, 66–67; deorbiting, 30; descent, 28–30, 33–34; docking, 189 (ill.); electrical power for, 83; environmental control system, 73; exercise on, 77; extravehicular mobility unit, 71 (ill.), 73; food on, 79–83; fuel for, 23, 25; fuel loading, 20; fuel tank, separation from orbiter, 25–26; galley, 82–83; hatch, 69; heat shield, 30; housekeeping on, 77; hygiene on, 79; landing, 31–34; launch, 20; liftoff, 21–23; manipulator arm, 84; manned ma-

neuvering unit, 72 (ill.); meals on, 79–83; mission station, 74; orbital maneuvering system, 27–30; payload handling station, 73–74, 74 (ill.); pets on, 77; prelaunch, 20; power module, 138, 139 (ill.), 140, 176 (ill.); recreation on, 76; reentry, 30–31, 34, 59; rendezvous and docking station, 73–74; rescue system, 69 (ill.); sleeping facilities on, 77, 79; solid rocket boosters, 24–25, 33; thermal protection system, 48, 59 (ill.), 59–60; Velcro materials on, 76; windows, 67
Orbiter Vehicle 101; *see* Shuttle

Pallets, 145, 146 (ill.), 152 (ill.)
Parker, Dr. Robert A., 114
Passengers, on shuttle, 118
Patrick Air Force Base, 86
Payload: capacities of, 89; cost of, 120, 122; handling of, 74; operations, 75 (ill.); transfer, 94–96, 98 (ill.); types of, 101, 151, 121–22, 124–25, 127
Payload changeout room, Kennedy Space Center, 96, 98 (ill.)
Payload handling station, on shuttle orbiter, 73–74, 74 (ill.)
Planetary vehicle, manned, 226–27 (ill.)
Power module, 138, 139 (ill.), 140, 192
Prelaunch procedures, 20

Reaction control system (RCS); engines, 56; location of, 56; purpose of, 56
Recovery and Disassembly Facility, Kennedy Space Center, 101
Recreation, on shuttle orbiter, 76
Rectenna, solar-power satellite, 201
Reentry, 30–31, 34; alignment with runway, 31; altitude at, 30; heat shield, 30; speed, 33; temperature, of hull, 33; temperature, of vehicle surface, 30
Rendezvous and docking station, on shuttle orbiter, 73–74
Rescue system, on shuttle orbiter, 69 (ill.)
Rocket port, first, 86
Rockets, early designs, 38–44
Rosen, Milton W., 35

Rover, lunar, 216, 218 (ill.), 221 (ill.)
Ryan, Cornelius, 42

Sagan, Carl, 107
Salyut, 154, 172, 187
Sänger, Eugen, 42
Search for extraterrestrial intelligence system (SETI), 192; antennas for, 191 (ill.), 194 (ill.)
Service and Access Tower, Kennedy Space Center, 97 (ill.)
Shuttle: altitude of, 48; components of, 53 (ill.); crew, number of, 48; designs of, 44, 48; designs, theoretical, 40–42, 44, 46; external tanks, 49 (ill.), 51, 52–53 (ill.), 54, 55 (ill.), 178 (ill.); flight date, 43, 46, 48; fuel consumption of, 54; interim upper stage, 130, 131–32 (ill.), 133, 133 (ill.), 135; main engines, 54; manipulator arm, 131 (ill.); orbital test flights, 48; power module, 138, 139 (ill.), 140; proper name of, 48; separation motors, 50; shuttle orbiter, 49 (ill.), 58 (ill.), 64–65 (ill.); solar electronic propulsion system, 134 (ill.), 136–137; solid rocket boosters, 24–25, 49 (ill.), 50–52, 102 (ill.); solid spinning upper stage, 135 (ill.), 137 (ill.), 137–38; teleoperator retrieval system, 129 (ill.), 129–30
Shuttle orbiter; *see* Orbiter, shuttle
Shuttle/Tethered Satellite system, 192
Signal Communication by Orbiting Relay Equipment (SCORE), 182
Skylab, 19, 41, 77, 79, 81–83, 96, 106, 110, 124, 129–30, 151, 154, 172, 187; spider on, 77
Slayton, Donald K., 109
Sleeping facilities, on shuttle orbiter, 77, 79
Solar array wing, 186 (ill.)
Solar electronic propulsion (SEP), on shuttle, 134 (ill.), 136–37
Solar-power satellites (Sunsats), 197–210, 202 (ill.), 204 (ill.); automated beam builder, 205 (ill.), 208; capacity of, 207; cargo delivery, 206 (ill.); central core, 200 (ill.); concept of, 198; cost of, 198, 201; design of, 198; electrical energy for, 201, 203; funds for, 207; launch date, estimated, 201; microwave beam, 207–8; microwave conversion, 198, 203; rec-

tenna, 201; solar (photovoltaic) cells, 203–5, 207; solar collectors, 201

Solid rocket boosters (SRBs), 49 (ill.); altitude at separation, 24; descent, method of, 52; nose cone of, 51; recovery of, 102 (ill.); separation from shuttle, 24–25, 50–51

Solid spinning upper stage (SSUS), 135 (ill.), 137 (ill.), 137–38

"Sortie Can," 148

Space colonies, 222–24; Bernal sphere design for, 223 (ill.); concept of, 222; sizes of, 224

Space construction base; *see* Construction base

Space exploration, in literature, 35–36

Spacelab, 141–56; airlock, top, 151; airlock tunnel, 150 (ill.); atmosphere, magnetosphere, and plasmas in space, 144 (ill.), 151, 155 (ill.); concept of, 145; interior of, 149; layout of, 149 (ill.); long modules, 148; materials sciences laboratory, 153; missions, forms of, 148; mockup, 147 (ill.); optical window and viewport, 151; in orbit, 142–43 (ill.); preliminary concept of, 148 (ill.); temperature of, 151

Spacelab projects: advanced technology laboratory, 153; astrophysics payload, 151; life sciences-biology, 153; solar physics payload, 151

Space Science Board, National Academy of Science, 158

Space shuttle; *see* Shuttle

Space Shuttle Flight 7 (SS-7), proposed first operational flight, 46

Space shuttle main engines (SSMEs), 55, 57 (ill.); fuel consumption of, 55; flight record, purpose of, 56

Space technology, early theories of, 39

Space telescope, 101, 157–70, 159 (ill.), 161–65 (ill.); capabilities of, 162–63; components of, 164; concepts of, 158; dimensions of, 160; guidance system of, 160; launch date, 160; release from payload bay, 161 (ill.); servicing of, 162 (ill.); uses of, 164

Space stations, 171–80, 174–75 (ill.); crews for, 172; future systems, 177; lunar, manned, 212, 214–15 (ill.); possibilities for, 172–73; proposals for, 172; space construction base, concepts of, 173, 176, 179

Space transportation system, concepts of, 45 (ill.)

spider, on *Skylab*, 77

Spielberg, Steven, 121

Sputnik, 180, 182

"Star Trek" *Enterprise*, 43, 48, 76

Stations; *see* Mission station; Space stations

Structural Test Article 099 (STA-099), 64

Sudbury Astrobleme, 224

Suits, space, 117 (ill.)

Synchronous Communication Satellite, 101

Teleoperator retrieval system (TRS), 129 (ill.), 129–30

Telescope; *see* Space Telescope

Thermal protection system, of shuttle orbiter, 48, 59, 59 (ill.), 60

Tiros weather satellites, 86, 124

Titan, 44; Titan III, 44, 132; Titan 34-D, 132; Titan *Centaur*, 44

Tracking and Delay Relay satellite, 101, 103

Tsiolkovsky, Konstantin Eduardovich, 37–38

2001—A Space Odyssey, 42, 172, 177

V-2 rockets, 39, 43, 86

Vandenburg Air Force Base, 31, 88–89

Vehicle Assembly Building, Kennedy Space Center, 94, 96

Vehicles: lunar rover, 218 (ill.), 221 (ill.); orbital transfer vehicle, 177, 179

Velcro materials, 76, 82

Venus, 225, 228; mission to, 225

Verne, Jules, 36

Vertical Processing Facility, Kennedy Space Center, 101

Water-mining and processing station, lunar, 216 (ill.), 217

Weather satellites, 86, 124, 182, 184

Wells, H. G., 36

Whipple, Fred L., 42

White Sands, New Mexico, 86, 103

Wright brothers, 192

Wright, Wilbur, 47

Yardley, John F., 62

Young, John, 109

ROBERT M. POWERS, author of Stackpole's *Planetary Encounters* (1978), has been writing about science and space for over fifteen years and has more than three hundred articles to his credit in magazines, periodicals, and newspapers, including *Saturday Review, Harper's,* and the *Los Angeles Times*. He has covered launches from Kennedy Space Center, as well as the Viking landings on Mars at the Jet Propulsion Laboratory. A Fellow of the Royal Astronomical Society and Member of the American Association for the Advancement of Science, Powers also serves on the board of trustees of a major metropolitan science center. Mr. Powers currently resides in Denver, Colorado.